Programming
the Parallel Port

Interfacing the PC for Data Acquisition and Process Control

Dhananjay V. Gadre

CRC Press
Taylor & Francis Group
Boca Raton London New York

CRC Press is an imprint of the
Taylor & Francis Group, an **informa** business

CRC Press
Taylor & Francis Group
6000 Broken Sound Parkway NW, Suite 300
Boca Raton, FL 33487-2742

First issued in hardback 2017

© 1998 Taylor & Francis.
CRC Press is an imprint of Taylor & Francis Group, an Informa business

No claim to original U.S. Government works

ISBN 13: 978-1-138-41256-9 (hbk)
ISBN 13: 978-0-87930-513-0 (pbk)

**Visit the Taylor & Francis Web site at
http://www.taylorandfrancis.com**

**and the CRC Press Web site at
http://www.crcpress.com**

To Chaitanya and Sangeeta

Foreword

No other interface has been so constant since the PC was introduced in 1981. Originally implemented to provide a "high speed" interface to the latest generation of dot matrix and daisy wheel printers, the parallel port has become the most common interface used to connect a wide variety of peripherals.

For many years, up until around 1989, printers were the only peripheral that took advantage of the parallel port. The port was viewed primarily as a "printer" port and other types of peripherals did not use it. Then companies such as Microsolutions and Xircom got the idea that you could actually use the port to get information back into the computer, and therefore use it as a bi-directional communication port. Being parallel, you could get much higher performance than using the PC's serial port, with greater simplicity.

The old parallel port became an easy-to-use interface for connecting peripherals. With a very simple register model, it is easy to get information into and out of the PC. The only drawback was that it was relatively slow. The CPU and platform performance was increasing at a tremendous rate, but the I/O capability of the PC stayed the same. While the CPU increased 100 fold, the parallel port remained stagnant.

This all changed with the formation of the IEEE 1284 Committee in 1992. This committee, sponsored by the Institute of Electrical and Electronic Engineers, had the charter to develop new, advanced parallel port modes that would enable high speed bi-directional data transfer through the parallel port. The requirements was to do this and still be 100% compatible with "standard" parallel port. Working with industry groups and individuals, the IEEE 1284 committee produced its new standard in 1994. This standard, IEEE St. 1284-1994, defined new ways of using the parallel port for high speed communication.

Two of these new modes are the EPP and ECP modes. Now, rather than being limited to a software-intensive, 50Kb-per-second port, you can get simple data transfer at rates approaching 2Mb per second. This 40 fold improvement in throughput is even more remarkable considering that the modes also remain backwards compatible with existing devices and interfaces.

This standard has enabled a wide range of peripherals that take advantage of the parallel port. Almost all new peripherals provide support via the parallel port. This includes the traditional uses such as printers, scanners, CD-ROM, hard drive, port sharing, and tape, as well as some non-traditional uses.

One of the most popular, non-traditional uses of the 1284 parallel port has been as a scientific and data acquisition interface. The past few years has seen tremendous growth in the use of this port for attaching control devices and for use as a simple interface for data acquisition instruments. The ability to have the same PC interface in the lab and on every portable computer makes this the ideal port to attach this type of equipment.

In this book, *Interfacing to the PC using the Parallel Port*, Dhananjay provides a clear introduction and model on how to use the parallel port for these types of applications. This is the ideal reference book for anyone wishing to use the PC for interfacing to external devices. Dhananjay presents a step-by-step approach to the subject. Starting with the basic, "What is the Parallel Port?" and "What is Data Acquisition", he leads you up the path to designing peripheral interfaces and writing the software drivers necessary to control and communicate with your devices.

I'm sure you'll find this an invaluable tool in aiding your understanding of the parallel port and the concepts and implementations of data acquisition peripherals.

Larry A. Stein

Larry Stein is the Chair of IEEE 1284.3 and 1284.4 Committees. He was instrumental in the development of the IEE 1284 standard and served as chair of the EPP Committee. He is currently Vice-President of Warp Nine Engineering and is the chief architect of the Warp Nine interface cards and IEEE 1284 Peripheral Interface Controller.

Acknowledgments

My interest in parallel printer adapters began in 19980 when Professor Vijaya Shankar Varma at the Delhi University asked me if I could build a resistor DAC for the parallel port. Since then, together with Dr. Pramod Upadhyay, we have enjoyed building and using many devices on the parallel port. It has been a pleasure working with them. Most of these gadgets were built at the Centre for Science Education and Communication (CSEC), University of Delhi.

While we were at it, Professor Pramod Srivastava, Director of CSEC, was a constant source of suggestions and useful comments. He was an even bigger help in providing financial support for the projects.

Since coming to the Inter-University Centre for Astronomy and Astrophysics (IUCAA) in Pune, India, Pravin Chordia has been a great help in building many of the devices. Arvind Paranjpye suggested the photometer interface problem, which was completed as another project. Manjiri Deshpande provided useful suggestions and evaluated some of the ideas presented here.

Professor S.N. Tandon, my boss at the Instrumentation Laboratory, allowed me to use the facilities in the laboratory for building many of the projects described here. Working with him has been an education for me and I thank him for many of the things I learned from him.

I learned the finer points of UNIX and Linux from Sunu Engineer. A brilliant programmer that he is, all the Linux-related projects would have been incomplete without his collaboration. He also read through many of the chapters in this manuscript and provided critical comments.

This work has been possible, in no small measure, because of the atmosphere of academic freedom I enjoy at IUCAA, and I thank Professor J.V. Narlikar, Director of IUCAA, for creating this wonderful place and providing me with a chance to work here.

Thanks are due to Dr. James Matey, Contributing Editor of *Computers in Physics*; to Joan Lynch, Managing Editor of *EDN*; to Jon Erickson, Editor-in-Chief of *Dr. Dobb's Journal*; and to Lindsey Vereen, Editor-in-Chief of *Embedded Systems Programming*; for providing me the opportunities to write for their respective journals.

I thank Jon Erickson (*DDJ*), Mike Markowitz (*EDN*), and Lindsey Vereen (*ESP*), for allowing me to use the material from their respective journals for this book.

Larry Stein, of Warp Nine Engineering and Chairman of IEEE's P1284 committee, was a great help in providing details about the EPP and ECP, and I thank him for that.

Thanks are also due to Santosh Khadilkar for his help in organizing the manuscript for this book. This manuscript was prepared using the IUCAA computer centre facilities.

I am delighted to thank my wife, Sangeeta, for her encouragement and her patience. She fought like a lone warrior in engaging and containing our son, Chaitanya, while I was busy. It was only because of her support that this work could be undertaken, and I cannot thank her enough.

This acknowledgment would be incomplete without placing on record my deep sense of gratitude to the foresight of my parents, Aai and Nana, in providing me a decent education even in the face of severe financial crunch. I think nobody else would be happier than Aai and Nana in seeing this book in print.

Dhananjay V. Gadre
Pune, India

Dhananjay Gadre is a Scientific Officer with the Instrumentation Programme of the Inter-University Centre for Astronomy and Astrophysics, Pune, India. He has been working with the IUCAA for the past four years. Previously, he was a lecturer at the SGTB Khalsa College, University of Delhi, teaching undergraduate electronics for about four years. He is now a graduate student at the Microelectronics Research and Communications Institute, Electrical Engineering Department, University of Idaho, on study leave from IUCAA.

Table of Contents

Chapter 1

Introduction

Data acquisition is the process of gathering information on a phenomenon by acquiring data on one or more variables. An example of a data acquisition process is recording the variation of ambient temperature as a function of time. For automated data acquisition, you need suitable sensors and associated hardware that can connect the sensor(s) to a host computer. You also need the software necessary to transport and translate the data from the sensor(s) to the host. This book is not about sensors or associated hardware, but it is about ways you can connect a sensor and its hardware to a PC using an efficient and unconventional interface: the parallel port.

Why the Parallel Port?

Conventional methods for connecting external hardware to a PC include the use of plug-in interface cards. This approach has several disadvantages, such as:

- If the device is meant for lab or classroom use, placing hardware inside the computer may be too risky for the machine or the users (who could be beginners). A piece of hardware is easily accessible for probing and measuring when it is outside the confines of a PC. Inserting an interface card increases the complexity of the operation. In some cases, adding an interface card could be a recipe for disaster (for instance, when you're interfacing to a multimeter or logic analyzer or an oscilloscope probe that may create unwelcome electrical shorts).

- Not all computers have an available expansion slot. With shrinking computer sizes, some modern computers have fewer slots. Laptop computers do not have any conventional expansion slots (other than PCMCIA slots). Other computers may have slots, but those slots may be devoted to other purposes, such as network cards, sound cards, and fax/modems.

- Many applications that require data acquisition and control do not really require the sophistication of a motherboard expansion slot. A simpler solution would be cleaner, easier, and cheaper.

An alternative to using an interface card is to design your hardware so that it can connect to the PC through the parallel printer adapter (i.e., the parallel port). Parallel ports are universally available on all PCs and compatibles. Another benefit of the parallel port is that the IEEE has continued to improve the parallel port specification while at the same time retaining backward compatibility with the original parallel port. Over the past few years, programmers have increasingly favored the parallel port as a means of connecting tape backup systems, CD-ROM Players, and LAN adapters, as well as various types of high-performance printers.

The parallel port is thus an elegant solution for interfacing a data acquisition device with a PC, and this book will show you how to do it. The last chapter of the book shows how to interface the parallel port under the Linux OS. The schemes described in this book are not the only or even the best methods for implementing data acquisition in every situation. The code in this book is written primarily from a DOS perspective. I have not sought out the higher end nuances of Windows programming, such as device drivers and Win32 API calls. My purpose is to present inexpensive alternatives for data acquisition and to provide a basic understanding that each reader can then adapt to specific tasks.

What is Data Acquisition?

Data acquisition is the process of acquiring information about a phenomenon. If you are studying a variation in ambient temperature with time, your data acquisition could consist of measuring and recording the temperature either continuously or at some discrete interval. An automated, human-readable data acquisition system for this situation would employ a suitable temperature sensor (e.g., a thermister) connected to a strip-chart recorder. The strip-chart recorder would move the paper in one direction at some rate, and a stylus driven by the sensor output would plot the temperature in the orthogonal direction, thereby creating a continuous record of the temperature–time variation. A computerized solution for this scenario would essentially do the same thing, except that instead of writing the data to a strip-chart, the sensor and its associated components would transmit the data through some hardware interface to the PC. A computer running a suitable software package (the data acquisition program) can acquire, display, process, and store the data. The advantage of using a computer for data acquisition is that a computer has the flexibility to adapt to changing needs and to further process the resulting data to enhance its usefulness.

Figure 1.1 shows the block diagram of a simple computer-assisted data acquisition system. A computer is connected to the interface hardware. The interface hardware, in turn, is connected to suitable sensors that will respond to changes in the physical variables for the experiment.

Control is the process of acquiring data about a phenomenon as a function of some variable and then regulating the phenomenon by restricting the variable to a preset value. For instance, if you wanted to control the temperature of a furnace, you would need data acquisition hardware as in Figure 1.1, as well as additional hardware to control a heater heating the furnace. The data acquisition hardware would measure the furnace temperature (see the sidebar "Trivial Pursuits"), which would then be compared with the required (preset) value. If the temperature is not equal to the required value, a corrective action would occur. Figure 1.2 shows the block diagram of a computer-assisted control system.

Intended Audience

This book is for anyone who is interested in using the PC for data acquisition or control. If you are developing data acquisition hardware or instrumentation and looking for a smart way to interface that hardware to a PC, you'll find some answers in this book. Educators and hobbyists who are looking for simple, low-cost interface solutions will also find this book useful.

Figure 1.1 Block diagram of a typical automated data acquisition system.

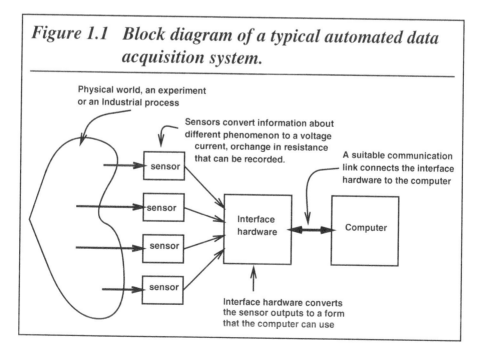

Organization of the Book

I will begin by describing the requirements for interfacing a computer to external control or data acquisition hardware. You will see that most of these requirements essentially boil down to providing an interface that has a suitable Analog to Digital Converter (ADC), Digital to Analog Converter (DAC), and digital latch for digital output or digital buffer for digital input.

Figure 1.2   Block diagram of a computer control system.

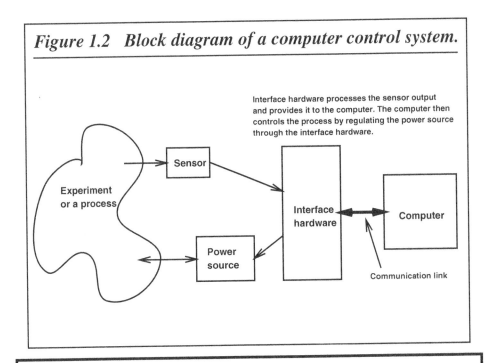

Interface hardware processes the sensor output and provides it to the computer. The computer then controls the process by regulating the power source through the interface hardware.

Experiment or a process

Sensor

Interface hardware

Computer

Power source

Communication link

Trivial Pursuits

It may seem impractical to use a computer just to control the temperature of a furnace, and in some cases, it is. However, for a system that requires very precise, high-quality temperature control, a computer may indeed be a practical solution. An oven or an ordinary home furnace uses a simple thermostat with an on–off control scheme to regulate temperature. This design results in considerable fluctuation up and down around the preset value. A computer could employ a more sophisticated control method, which would reduce fluctuations and achieve a closer agreement of the preset and the actual temperatures. Some typical control methods for this situation are the proportional, integral, or derivative methods. A computer is very well suited to implement such control schemes.

Before I describe how to interface these components to the PC, however, I will look closely at the interface connection. I will describe the parallel port in detail, describing the first parallel port interface and showing how the parallel port has evolved to keep pace with increasing PC performance. I will then show you a variety of ADC and DAC components that you can use in different environments. I will describe ways to perform digital input and output using the parallel port and how you can convert the PC into a virtual instrument by connecting a few more components to the parallel port. Subsequent chapters discuss a variety of development tools that will be of particular interest to microprocessor enthusiasts who are developing and building microprocessor-based applications (Figure 1.3).

Chapter 2 describes interfacing fundamentals and general requirements for building a computer interface. A good background in digital electronics will be very helpful for understanding this chapter, but it is not essential.

Chapter 3 discusses the history of the parallel printer adapter and describes the details of the standard parallel port.

Chapter 4 describes programming considerations for the parallel port. This chapter also describes the various ways of using the parallel adapter for simple applications.

Chapter 5 describes the Enhanced Parallel Port (EPP) and the Extended Communications Port (ECP).

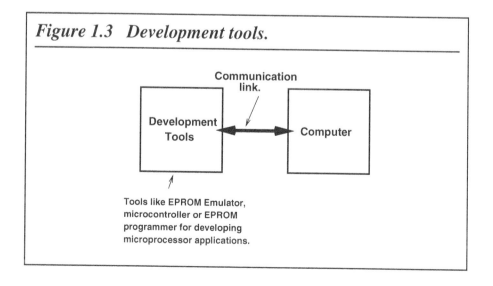

Figure 1.3   Development tools.

Chapter 6 looks at various Analog-to-Digital Converters (ADC) and Digital-to-Analog Converters (DAC) and shows how to interface these ADCs and DACs to a PC using the parallel port. Today, a wide variety of ADCs and DACs are available.

Chapter 7 shows how to build suitable hardware to measure the time period and frequency of digital signals. This chapter describes an interface for an astronomical photometer.

Chapter 8 presents a pair of complete data acquisition systems providing 8-bit and 12-bit resolution.

Chapter 9 describes how to add more bits to the parallel port.

Chapter 10 shows how to use of the parallel port to host an EPROM emulator. An EPROM emulator is a useful tool for testing microprocessor code for embedded applications.

Chapter 11 describes how to connect the parallel port to an external microprocessor. Two examples show how the parallel port can be connected to an ADSP-21xx-based circuit and to an AT89C2051 (an 8051 microcontroller variant) controller.

Chapter 12 describes how to use the parallel port to host an EPROM microcontroller programmer.

Chapter 13 discusses various ways to generate digital waveforms using the parallel port.

Chapter 14 discusses a Data Acquisition System (DAS) for the Linux operating system. An example application describes how this DAS can be used to collect and distribute data across a computer network. In the example, a weather station provides real-time weather data on the Internet using a World Wide Web (WWW) facility.

Companion Code on the FTP Site

NOTE: The full source code for this book is now available on the publisher's ftp site at `ftp://ftp.cmpbooks.com/pub/gadre.zip`; logon as "anonymous" and download the file. Any references to the "companion code disk" in the book now refer to the files available on the ftp site.

## Chapter 2

How to Build a Computer Interface

What is an Interface?

An interface is a system consisting of hardware, software, or both that allows two dissimilar components to interact. Consider, for example, the problem of connecting a special type of printer system manufactured on the planet Mars with a PC on the Earth. The manufacturer of the printer has provided complete specifications for the input signals, but these specifications unfortunately do not correspond to either the RS-232 port or the Centronics printer port attached to the PC. To interface this Martian printer with the earthbound PC, you must do two things. First, you must build suitable hardware that can connect the PC to the printer and generate all the signals required by this printer. The signals generated by the PC should meet the timing as well as the voltage level (or current level) requirements of the printer. Second, you must provide suitable software routines and drivers that will translate user commands such as `m_print file_name` into signals that the printer will understand.

Examples of Various Schemes for Data Acquisition

There are several ways to use a PC for acquiring data. The method you choose will depend on the following factors:

- the required acquisition rate, peak as well as average

- the nature of the data (for instance, whether it is in digital or analog form)
- the amount of data to be acquired
- whether the source of data communicates through a specific data transfer protocol

The answers to these initial questions will begin to suggest a design for your data acquisition system. You will start to see whether PC is fast enough to acquire data by polling the data source and whether the data can be acquired on the fly (or whether you will need to employ data buffers in case the peak data rate is more than the PC can handle). If the PC needs to process the data while it is being collected, the data acquisition scheme could have either an interrupt mechanism that interrupts the main program to signal the arrival of data, or a scheme with some kind of data buffer, or both.

Another important question is: what is the unit of this data? Does the data arrive as bytes or bits. If the data arrives as bits, we must assemble these bits into bytes.

If your PC is not fast enough, you must provide an intermediate hardware buffer to retain the data. If the PC must also analyze the data while it is coming in, you will also need an intermediate data buffer so that incoming data is not lost. In some situations, the PC cannot process the incoming data until all the data has arrived. In this case, you must provide a very large buffer to accommodate data for the whole exercise.

A Speech Digitizer

As an example of a data acquisition system, I will briefly describe a computer interface to digitize speech for a 10-second period. The idea is to build the necessary hardware and specify the software that will connect a microphone to a computer such that the computer can acquire a set of numbers that correspond to the voltage variations as detected by the microphone from the speech signal. Figure 2.1 shows the block diagram for the speech digitizer. The microphone converts the acoustic signals of the speech to a corresponding electrical signal. This signal is suitably amplified by the pre-amplifier. The pre-amplifier drives the waveform digitizer, which is nothing but an

---

Data Acquisition Methods

Broadly, there are two ways of designing the data acquisition software, the polled method and the interrupt method. The polled method requires that the user program check at regular intervals whether the data is available with the help of a construct called a 'flag'. The state of the flag determines whether data is available. The flag has two states '0' or '1'. A '0' state could imply data available and a '1' state could mean data not available. The flag is set up by the data source and after the program detects this state, the data is read and the flag is reset by the user program. The interrupt method of data acquisition requires that the data source 'interrupt' the user program through the interrupt scheme. The user program then suspends it's current operation and executes a special program called Interrupt SubRoutine (ISR) to read the data and to acknowledge to the source that the data has been read.

Analog to Digital Converter (ADC), which will be described in some detail in a subsequent chapter. The computer cannot handle an analog electrical signal, so you must use a digitizer to convert the electrical signal to a digital format. The waveform digitizer connects to the computer through a suitable digital circuit. The waveform digitizer, together with this digital circuit, forms the hardware interface. The hardware interface connects to the computer using a suitable communication path or link. (There are several methods for connecting the interface hardware to the computer.)

The block diagram in Figure 2.1 shows the interface link as a bidirectional link that is required to send conversion commands to the interface circuit. A conversion command from the computer will trigger the waveform digitizer to take an instantaneous sample of the speech signal and convert it to a number. After the conversion is over, the bidirectional link transmits the converted number back to the computer.

The software part of the speech digitizer interface is a program that:

- determines the sampling rate of the speech signal;

- acquires sufficient memory from the operating system of the computer to store the numbers corresponding to a 10-second speech recording;

- issues a waveform convert command at the required rate;

- reads back the converted number; and

- stores the numbers in a file at the end of the record period.

At this stage, I have a rough design of the microphone interface. I must now address two important issues:

- the structure of the *digital circuit* that connects the digitizer to the communication link and

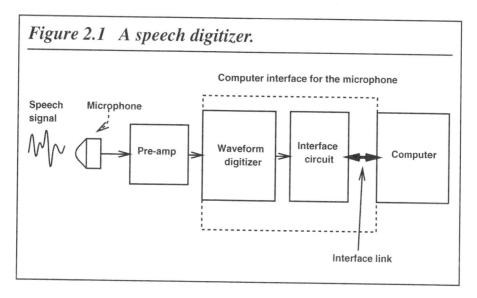

Figure 2.1 A speech digitizer.

- the communication link itself.

The digital circuit could be of many types and would depend, to some extent, on the choice of the communication link. The choice of a communication link also depends on the digital circuit, so the question becomes a sort of a chicken and egg problem. Options for the digital circuit include:

- The digital circuit could be designed such that it receives a command from the computer program to initiate a conversion of the digitizer circuit. The digital circuit triggers the digitizer and gets the converted number. The digital circuit then informs the program that the conversion is over and the converted number is ready. The program then reads the digital circuit and gets the number. The digital circuit then waits until the program sends a trigger command for a fresh conversion. The data output of the digitizer is presented by the digital circuit in parallel format (i.e., all the digital bits representing the number are available at the same time). This scheme means that the communication link must be able to handle data transfer rates of at least 10,000 conversions/s, assuming that the speech is to be sampled at 10,000 samples/s.

- You could design the digital circuit to hold all the data for a 10-second sampling period and transmit the data at the end of the period. For a 10-second recording at the rate of 10,000 samples/s, the digital circuit would need 100,000 memory locations. The computer program triggers the data acquisition process, and the PC is then free to execute other tasks. The digital circuit in the meantime performs 100,000 conversions, stores them temporarily in its internal memory, and at the end of the recording period, informs the program that the recording is over and that the data can be read back by the program from the circuit's internal memory. The computer program then reads out the memory contents of the digital circuit. With this method, because the communication link is not involved in data transfer in real time, the speed requirement of the link could be rather low.

- In a variant of the first method, the data could be transferred between the digital circuit and the computer in serial format. This requires only a few connections between the digital circuit and the computer; but at the same time, this method means the data transfer rate must support at least 100,000 bits/s (assuming each converted number can be represented by a 10-bit number).

The communication link could be one of the following:

- the RS-232 serial port
- the Centronics parallel printer adapter
- one of the many motherboard buses: the ISA, EISA, or PCI
- the SCSI bus that is available on many PCs and all Macs or the Universal Serial Bus (USB) on newer PCs.

The RS-232 serial port on the PC offers standard data transfer rates of up to 19200 baud, which translates to a maximum of about 1,900 bytes/s. Enterprising programmers, however, can program the RS-232 circuit to transmit and receive at 110,000 baud, which is about 10,000 bytes/s. The choice of the RS-232 port would put an additional burden on the digital circuit, in the form of a corresponding signal translator (the RS-232 protocol uses unconventional voltage levels to encode the low and high-level signals), as well as an RS-232 communications controller, which translates the serial RS-232 data to parallel format. The RS-232 interface does not offer any power supply voltages, and the interface circuit would need to have its own suitable power supply.

The use of a motherboard bus (ISA, EISA, or PCI) would allow data transfers at the fastest rates of all methods, in the range of 2,000,000 bytes/s and more. The motherboard requires special PCBs with edge connectors to connect into the motherboard slots and a relatively more complex digital circuit than the printer adapter solution. The motherboard slots, however, offer all the power supply voltages that the PC uses (+5, +12, –12, and –5V) to the interface circuit.

The SCSI bus, as well as the USB, can handle data transfer rates required by the microphone interface but are relatively more complex in comparison to all the above methods. Among these communication link choices, the parallel printer port clearly offers an inviting combination of speed and simplicity.

Data Acquisition for a CCD Camera

As a second example, let me describe a project I am currently working on: designing a controller for a Charge Coupled Device (CCD) camera and data acquisition system based on the PC.

This CCD camera problem doesn't pertain specifically to the parallel port, but I include it because it highlights some data buffering options that are important for both serial and parallel data acquisition.

A CCD camera is an electronic imaging device. The camera is composed of a CCD chip and associated electronics. The CCD chip converts incident light into packets of charge distributed over itself in small charge-trapping pockets called pixels. The associated electronics routes these charge packets to the output of the chip and converts the charge into voltage. The routing of the charge packet from a pixel to the output of the CCD chip is done using various clock signals called horizontal and vertical shift clocks. The electronics onboard the CCD controller generates these clock signals. Subsequently, the voltage corresponding to each pixel is digitized and sent to the PC. The CCD camera is controlled by the user from the PC and is connected to the PC through a suitable link. Most often, the link between the camera and the user's PC is a serial link, because in the case of this application (the camera will be used with a large telescope), the distance between the camera and the user's PC could be more than 20 feet and could even be a few hundred feet.

Figure 2.2 is a block diagram showing the CCD camera system. The user defines the format of the image through a data acquisition program running on the PC. This PC program then transfers the image parameters to the CCD controller and waits for the controller to send the image. (The actual process is more involved than this, but this description is sufficient for the purposes of the present discussion.)

The CCD controller triggers the CCD chip (as per the user parameters) and encodes the CCD pixel voltage (which is analog) into a digital number that can be handled by the user PC. For high-performance CCD cameras, the controller is equipped with a 16-bit ADC, so the data for each pixel consists of two bytes. The pixel data, now encoded as a number, needs to be sent to the user PC over the serial link. The user PC needs to be ready to receive the image pixel data and to display, process, and store the image. The CCD controller sends each pixel data as two bytes, one after the other. The time between two bytes of the pixel is *T1* and that between two pixels is *T2*. To get the image in minimum time, we need to minimize *T1 + T2*.

Depending upon various constraints, a number of different options for the CCD camera system will emerge. The constraints are nothing but pure economics:

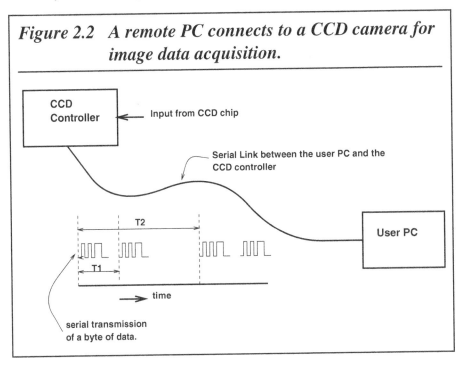

Figure 2.2   A remote PC connects to a CCD camera for image data acquisition.

- How do you get the image into the user PC at a minimum of hardware cost with the greatest possible elegance and ease of operation?
- How do you make the camera easily serviceable and easily upgradable?

Some possible solutions follow.

Case 1 The user PC needs to get the image as fast as the controller can send it. The user PC cannot wait to receive each and every byte of the image, because the user PC's primary task is to view and analyze the images.

Solution To minimize the image acquisition time, you must minimize $T1$ and $T2$. Also, because the user program cannot receive each and every byte of the pixel, one possible solution is to employ image buffer hardware in the user PC. The image buffer is nothing but read/write memory of sufficient size to store the incoming image. For instance, if the CCD chip is 1,024 rows with 1,024 pixels in each row (a typical case), then the total number of pixels is roughly one million (1,048,576) pixels. At two bytes/pixel, the required memory buffer would be about two Mb. The incoming image would be stored in this image buffer, and at the end of image acquisition, the user program would be informed. The user program would then transfer the image from the image buffer into the internal memory of the PC and free the image buffer memory to prepare for the next image. The effective time for the PC program to actually acquire the image is the time taken by the image buffer to acquire the image plus the time taken by the PC to transfer the image from the image buffer to internal memory.

Case 2 The user PC needs to acquire the image in the shortest possible time, but the user cannot afford full-image buffer memory.

Solution For this case, because the constraint is memory available in the PC data acquisition system, the solution lies in using a small memory buffer that is partitioned into two parts. At any given time, the incoming data is routed to one section of the buffer. When this buffer is full, an interrupt is generated, and at the same time, the incoming data is routed into the second section of the buffer. The interrupt is used to signal to the user program that the first section of the buffer is full and the contents should be transferred into the system memory. The user program then executes an ISR, which transfers the contents of the first buffer into the PC's internal memory. The data acquisition circuit, in the meantime, is still transferring the incoming data into the second buffer. When the second buffer is full, it will again generate an interrupt and start transferring the incoming data into the first buffer. This process will cycle so that incoming data is temporarily stored in the buffer before it is transferred into the main memory. An important requirement while implementing this solution is that the average rate of incoming data should be substantially less than the average rate at which the PC can transfer data between the buffer memory and the internal memory. Otherwise this solution cannot be implemented.

Case 3   The PC cannot afford buffer memory and must keep the acquisition hardware cost to a bare minimum.

Solution   The incoming data is stored in a latch. (A latch is a device that stores one data value. We will talk more about such devices later in this chapter.) A flag is set up to indicate the arrival of the data. The user program is continually monitoring the state of this flag and as soon as the flag is set, the user program reads the data latch, resets the flag, stores the data in internal memory, and again starts polling the flag. This continues until the PC receives the entire image.

Signal and Timing Diagram Conventions

In this book, I will adopt the conventions used in Adam Osborne's and Gerry Kane's classic *Osborne 16-Bit Microprocessor Handbook*. One issue of importance in digital circuits is the active level of the signal. Because the digital signal can have two levels (actually three, but I'll discount the third level at the moment), it is useful to define

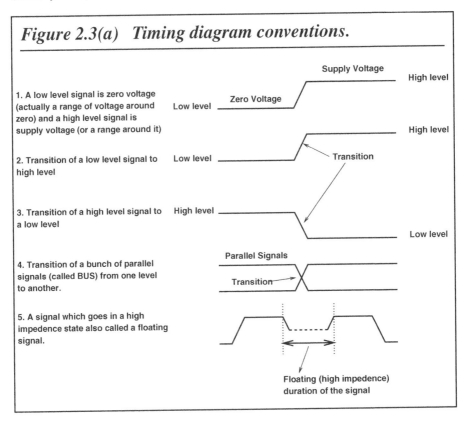

Figure 2.3(a)   Timing diagram conventions.

1. A low level signal is zero voltage (actually a range of voltage around zero) and a high level signal is supply voltage (or a range around it)

Supply Voltage
High level
Low level   Zero Voltage

2. Transition of a low level signal to high level

High level
Low level   Transition

3. Transition of a high level signal to a low level

High level
Low level

4. Transition of a bunch of parallel signals (called BUS) from one level to another.

Parallel Signals
Transition

5. A signal which goes in a high impedence state also called a floating signal.

Floating (high impedence) duration of the signal

which level is active. Active low signals are shown with a bar or an asterisk ($\overline{\text{WR}}$ or WR*) wheras active high signals do not have a bar or a star. Figures 2.3(a)–(c) show the timing diagram conventions for this book.

Hardware Components

So far, I have discussed some basic data acquisition problems and provided possible solutions at a very preliminary level. In this section, I will describe some of the hardware components used in real data acquisition systems.

Digital IC Families

The TTL family of digital ICs is one the most popular digital ICs. The 74xxx was the first of the TTL family. Since then many improvements in device processes and fabrication technologies have led to the introduction of more families offering improved performance over the standard 74xxx family. The various subfamilies in this series offer high speed of operation, low power dissipation, robust performance, and wide availability. The various subfamilies are 74LS, 74ALS, 74S, and 74F series.

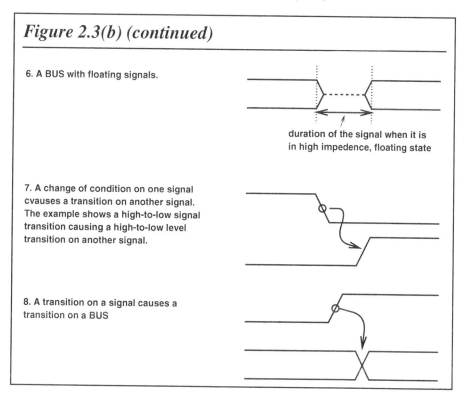

Figure 2.3(b) (continued)

6. A BUS with floating signals.

duration of the signal when it is in high impedence, floating state

7. A change of condition on one signal cvauses a transition on another signal. The example shows a high-to-low signal transition causing a high-to-low level transition on another signal.

8. A transition on a signal causes a transition on a BUS

The CMOS family is another important family of ICs. The 4000 series from Fairchild was the original member of the CMOS family. The components of this family offer very low power dissipation and wide supply voltage operation compared to the TTL. The sub families are 74HC, 74HCT and 74C.

Logic Levels and Noise Margins

Digital components need a supply voltage to operate. The voltage levels at input and output are related to the supply voltage levels. It may seem that if the digital circuit operates at +10V, the logic low is 0V and logic high is +10V. This is not so. A range of voltages around the two supply levels (0 and +10V) qualifies as a valid logic low and logic high.

Take the case of a low-power TTL component. This component operates at +5V supply voltage. The specifications require that, for error-free operation, an input voltage of up to 0.8V qualifies as logic low. Thus, an input voltage between 0 and 0.8V qualifies as logic low. An input voltage with a minimum of 2.0V qualifies as logic high. This means that an input voltage between 2.0 and 5.0V qualifies as logic high.

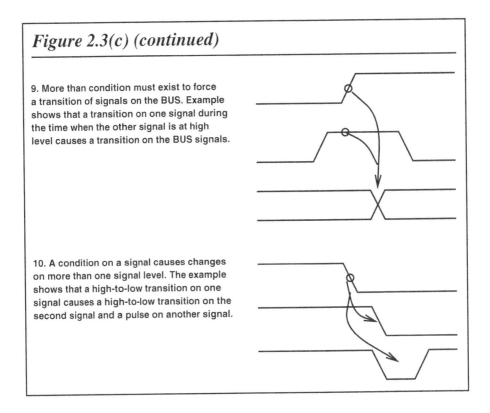

Figure 2.3(c) (continued)

9. More than condition must exist to force a transition of signals on the BUS. Example shows that a transition on one signal during the time when the other signal is at high level causes a transition on the BUS signals.

10. A condition on a signal causes changes on more than one signal level. The example shows that a high-to-low transition on one signal causes a high-to-low transition on the second signal and a pulse on another signal.

The low-power TTL specification guarantees that the maximum logic low output of the device will be 0.4V and a minimum logic high voltage will be 2.4V. These are called the worst-case output levels of the device. (These worst-case figures assume certain load conditions.)

Noise margin is defined as the difference in the voltage levels (for a given logic) of the input and output of a device. The maximum acceptable input voltage level for logic low is 0.8V. The maximum output voltage level for logic low is 0.4V, so the noise margin for logic low is 0.4V (Table 2.1).

For the high-level noise margin, you must consider the input and output voltages at the high end of the range. A minimum voltage input of 2.0V qualifies as logic high. The device would generate a minimum output voltage of 2.4V for logic high. The difference is 0.4V. So for LSTTL components, the noise margin is 0.4V.

To understand the purpose of the noise margin parameter, consider the case of two components of the same family, with the output of one device driving the input of the other device. The output of the first device is guaranteed to be less than 0.4V for logic low output. This voltage is connected to the input of the second device. I'll assume some noise gets added to the output voltage of the first device. How much of this noise can be tolerated if the second device is still to regard the voltage as logic low? Because the device can allow a maximum of 0.8V as logic low, the noise can be a minimum of 0.4V. This is the noise margin.

Noise margin figures vary from family to family. For the noise margin of a particular device, consult the data sheet for the device.

TTL and Variants

The circuit for the standard TTL NAND gate in Figure 2.4 shows a multi-emitter input transistor (transistor Q1) and an active pull-up output (transistor Q3) providing fast speed and low output impedance. Typical dissipations are 10mW per gate and a delay time (input to output) of 10ns. At the time these devices were introduced, this was revolutionary (fast speed and low power dissipation).

Table 2.1   Characteristics of various TTL series.

Type	Standard	S	LS	ALS	AS
Propagation delay time (ns)	10	3	7	4	1.5
Noise margin '0' (V)	.40	.30	.30	.40	.30
Noise margin '1' (V)	.40	.70	.70	.70	7.0
Power dissipation per gate (mW)	10	20	2	1	2
Fanout	10	10	10	10	10

Schottky (S) TTL

In this series of TTL gates, the transistors and diodes are replaced by Schottky transistors and diodes. A substantial decrease in propagation delay time is achieved as a result.

Low-Power Schottky (LS) TTL

This family offers combined advantages of low power dissipation and increased speed of operation.

Advanced Schottky (AS) TTL

This series is a result of further development of the Schottky series of devices. These devices offer faster speeds (less propagation delay time) than the Schottky series at a much reduced power dissipation.

Figure 2.4   A TTL two-input NAND gate.

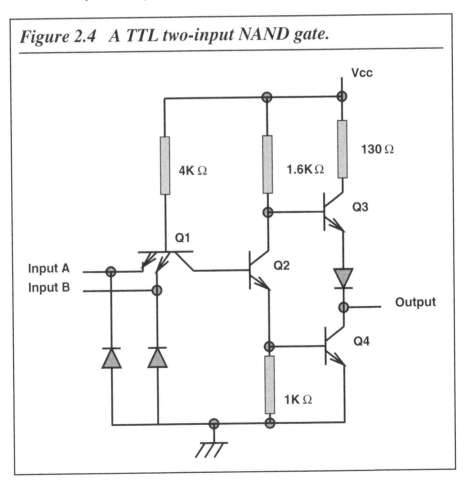

Advanced Low-Power Schottky (ALS) TTL

This series is a result of variations of the low-power Schottky series of devices. These devices offer faster speeds (less propagation delay time) comparable to the Schottky series but offer the lowest power dissipation.

CMOS and Variants

Besides the TTL components, the other popular digital component series use CMOS technology. The components of the CMOS family are the CMOS, HCMOS, and the HCMOS devices with TTL thresholds. The advantage of CMOS components is low power dissipation, wide operating voltage, and better noise immunity. These features make CMOS components very suitable for use in portable and battery-powered instruments.

Figure 2.5 shows a CMOS inverter. The CMOS inverter uses only two fets, Q1, a P-channel MOSFET and Q2, an N-channel MOSFET. When the input is low, the P-channel MOSFET conducts while the N-channel MOSFET is cut off. The output is a voltage equal to the supply voltage. When the input is high, the state of the MOSFETs reverse and the output is low.

At any time, either of two MOSFETs is cut off, hence the power dissipated by the device is extremely small. The only time the two MOSFETs conduct current is when the MOSFETs are switching. Therefore, the dissipation of CMOS components at DC is zero. Only when the switching frequency increases does the CMOS dissipate. At high frequencies, the power dissipation of CMOS components can equal or even exceed that of TTL components. The other disadvantage of CMOS components is the

Figure 2.5   A CMOS inverter gate.

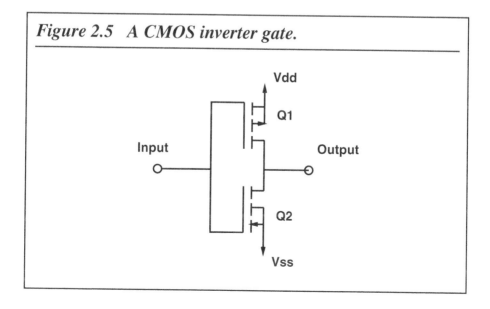

large propagation delay time. This prevents conventional CMOS components from operating at very high frequencies. The standard CMOS components are available in the CD40xxx series.

A variation of standard CMOS components is the HCMOS series. This family offers the high speed of Schottky devices and the low power dissipation of CMOS components. These components are available under the 74HCxxx series.

The problem with the HCMOS series is that its logic thresholds are incompatible with TTL components. It is not advisable to mix standard TTL and HCMOS components in a circuit, and the TTL output levels may not be recognized by the CMOS devices. So another variation of the CMOS, called the HCMOS with TTL thresholds (HCT), was introduced. HCMOS with TTL thresholds has all the advantages of HCMOS components, except the input logic thresholds were tuned to TTL levels. This series is known as the 74HCTxxx series. Most circuits described in this book use HCT components unless a particular component is not available.

Digital Components

The following sections discuss some contemporary digital ICs and their use in various circuits.

Sundry Gate

Almost any circuit may require digital gates. Gates that perform AND, OR, NOT, NAND, XOR, and XNOR Boolean functions are available in all the digital families cited above. Most popular are the gates from the HCT family for their low power dissipation and high frequency operation capabilities.

Table 2.2 lists some of the useful gates. You may wish to consider using logic gates with discrete components, such as a pair of switching diodes to make a two-input AND gate or a small signal transistor to make an inverter. In applications with board space crunch and where the application may allow, such gates can save board space as well as cost. Say you want a two-input AND gate. Rather than using a 14-pin DIP IC package for one gate, you can make this gate using a resistor and a pair of 1N4148-type switching diodes as shown in the Figure 2.6.

The Buffer IC and Some Cousins

Buffer ICs have the capacity to drive high current loads, which enable transmission of data at high speeds through signal cables with large capacitances. These devices also have significantly higher fanouts (more than 15) than ordinary gates. Normally the buffer IC has an output enable signal that can be used to control the flow of data to the output, which is very essential in a bus system. Typically the buffer IC has an active low output enable (referred to as OE*) pin. When the output enable signal is not active, the outputs of the device are in the high impedance state (also called the tristate). Many buffer ICs are also available with a control pin for each transmission element instead of a single control bit for a group of four gates. Another variation of the buffer IC is the bus transceiver IC. This device can transfer data bidirectionally and instead of an output enable control pin, it has a direction control pin. Table 2.3 shows the various buffer ICs.

Table 2.2   Some useful TTL gate ICs.

Name	Description
74HCT00	Quad, 2-input NAND gate
74HCT02	Quad, 2-input NOR gate
74HCT04	Hex, inverter
74HCT08	Quad, 2-input AND gate
74HCT10	Triple, 3-input NAND gate
74HCT14	Hex, Schmitt inverter
74HCT20	Dual, 4-input NAND gate
74HCT32	Quad, 2-input OR gate
74HCT86	Quad, 2-input XOR gate

Figure 2.6   A two-input AND gate realized using discrete components.

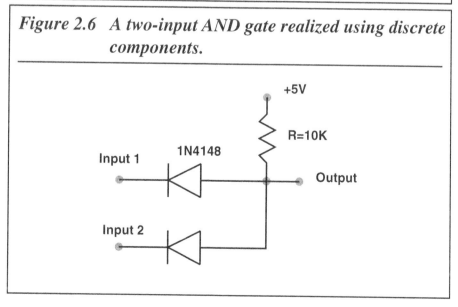

Flip-Flops and Latches

Flip-flops are ubiquitous devices important in any sequential circuit. Flip-flops are also called bistable multivibrators because they have two stable states. Flip-flops can be made out of discrete gates using suitable feedback. The various flip-flops are the D, S-R, J-K, and T types (Table 2.4). Flip-flops are used to remember a digital event and to divide clock frequencies, etc. A group of flip-flops in a package with a single clock input is called a latch.

Figure 2.7 shows the symbols of two of the flip-flops, shown with edge-triggered clocks. The clocks have their positive edge as the active edge.

Typically, latch ICs (Table 2.5) have a common clock input and an output enable control pin. The input data is transferred to the output at the rising (or whatever the particular case may be) edge of the clock. Data is actually available on the output pins only if the output enable control pin is active. Typically, a latch IC has a common latch enable pin and an output enable control pin. The data at the output follows the input as long as the latch enable pin is active. When the latch enable pin is inactive, the data at the output pins is frozen to their last state before the latch enable pin became inactive. The output data appears on the output pins only when the output enable control pin is in an active state.

Table 2.3   Some useful TTL buffer ICs.

Name	*Description*
74HCT240	Inverting, octal tristate buffer
74HCT244	Octal, tristate buffer
74HCT245	Octal tristate transceiver
74HCT125	Quad, tristate buffer with active low output enable
74HCT126	Quad, tristate buffer with active high output enable

Table 2.4   Some popular flip-flop ICs.

Name	*Description*
74HCT73	Dual, J-K flip-flop with clear input
74HCT74	Dual, D-type flip-flop with preset and clear
74HCT76	Dual J-K flip-flop with preset and clear
74HCT174	Hex D flip-flop with common clear and clock
74HCT273	Octal D flip-flop with clear
74HCT574	Octal D edge-triggered flip-flop with clear

The Decoder and the Multiplexer

Decoders play a vital role in interpreting encoded information. A decoder has n input lines and 2^n output lines (Table 2.6). Depending upon the state of the inputs, one of the 2^n output lines is active. The active state of the output may be '1' or '0'. Apart from the input and output lines, there may be some decoder control input lines. Unless the control input lines are active, none of the outputs of the decoder is active.

A multiplexer is a device that puts information from many input lines to one output line. This device has 1 output, n select, and 2^n input lines.

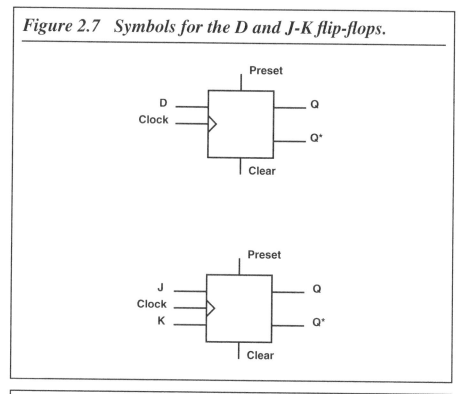

Figure 2.7   Symbols for the D and J-K flip-flops.

Table 2.5   Some popular latch ICs.

Name	Description
74HCT373	Octal D-type latch
74HCT573	Octal D-type latch

Counters

Counters, as the name suggests, count. The counter has a clock input pin, a reset input pin, and many output pins depending upon the type of the counter. An 8-bit binary counter would have eight output pins. After the reset pin is de-asserted, the outputs of the counters are all reset to '0'. Thereafter, at each pulse of the clock, the output of the counters would increase by one. It is not possible to go into the intricacies of counters, but a list of popular counter ICs is useful (Table 2.7).

Table 2.6   Some popular decoder and multiplexer ICs.

Name	*Description*
74HCT137	3-to-8-line decoder with address latch
74HCT138	3-to-8-line decoder
74HCT139	Dual 2-to-4-line decoder
74HCT251	Eight-channel tristate multiplexer

Table 2.7   Some popular counter ICs.

Name	*Description*
74LS90	Decade counter
74HCT4024	Six-stage binary counter
74HCT4040	12-Stage binary counter
74HCT190	Synchronous decade up/down counters with mode control
74HCT191	Synchronous hex up/down counters with mode control

Chapter 3

The Parallel Printer Adapter

The computer industry has at least four names for the parallel port: the parallel printer adapter, the Centronics adapter, the Centronics port, or quite simply, the parallel port. Any port that provides parallel output (as opposed to ports that provide data serially) is a parallel port, but in PC jargon, the term *parallel port* refers to ports conforming to a specification (and later enhancements) for what was originally known as the parallel printer adapter.

In the early days of personal computers, most printers only could be connected using a serial interfaces. When printers started to have their own memory buffers, users found the serial link too slow. Manufacturers started offering printers with a parallel interface that could, in principle, receive data at least eight to ten times faster than was possible with the serial port. The adapter on the computer that allowed the user to connect to the parallel printer (i.e., the printer with a parallel input) was the parallel printer adapter.

At the time the parallel printer adapter came into existence, PC processors were all 8-bit processors. So, it seemed logical to define a data path to a printer with the capacity to transfer eight simultaneous bits of data.

Anatomy of the Parallel Printer Port

The best way to understand the design of the parallel printer port is to work through the thought process of its original designers. The designers of the parallel printer port knew that:

- The port must provide eight data signals to transfer a byte of data in parallel.
- The computer must be able to signal to the printer that a byte of data is available on the data lines. This signal was called the strobe signal.
- The computer must get an acknowledgment signal from the printer. This signal is called the acknowledge (or ack) signal.

The data, strobe, and ack signals are sufficient to transfer data between the computer and the printer in a rather raw manner. More signals are needed to exchange more information between the computer and the printer. Printers are electromechanical devices with three primary tasks: to receive print data from the computer, to print this data, and to respond to user information (like changes in font, etc.). During occasions when the printer's internal memory buffer is full (because the printer cannot print as fast as it can receive data), the printer must be able to inform the computer that it can't receive more data or that it is busy. This signal from the printer to the computer is called the busy signal. The printer also needs to signal the computer if there is any error condition (e.g., if the paper has jammed in the printer mechanism or if the paper is empty). A signal between the printer and computer called the error signal is used for this purpose. The computer can also use more signals to control the printer, such as a signal line to reset the printer at the start of a fresh print run so that any residual data in the printer buffer is flushed out.

Table 3.1	**Signals of a hypothetical parallel printer adapter.**	
Signal	*Function*	*Source*
DATA (8)	transfer print data strobe to instruct the printer that new computer data is available	computer
strobe	to instruct the printer that new data is available	computer
acknowledge	acknowledgment from printer that data is received	printer
busy	indicates printer is busy	printer
error	indicate error condition on the printer	printer
reset	reset the printer	computer

The signals described in the preceding paragraph form a good foundation for a hypothetical parallel printer adapter. Table 3.1 shows the signals for this hypothetical port.

From Table 3.1, it is clear that to implement a parallel port, the computer actually needs three independent ports: an output port to transfer data to the printer, another output port to carry the strobe and reset signals, and an input port to read the acknowledge, busy, and error signals from the printer.

The actual parallel printer adapter is designed with 17 signals. These signals are distributed across the three internal ports as follows:

1. an output port with eight data signals called the DATA port;

2. an input port with five status signals called the STATUS port;

3. another output port with four signals called the CONTROL port.

The block diagram in Figure 3.1 shows the design of the parallel printer adapter. The PC system bus interface connects the adapter to the microprocessor signals. The output signals from the adapter are connected to a 25-pin D-type connector. On many

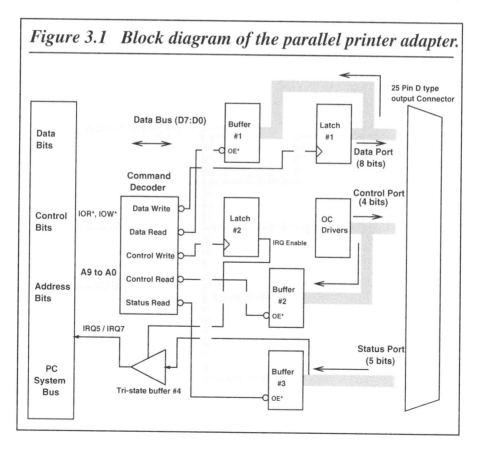

Figure 3.1   Block diagram of the parallel printer adapter.

of the new PCs, the parallel printer adapter has been integrated on the motherboard, though plug-in card adapters are also available.

The block labeled command decoder is nothing but an address decoder. The command decoder has address lines (A0–A9) as inputs, as well as the IOR* and IOW* CONTROL signals and five outputs labeled data write, data read, status read, control read, and control write. The data bits D0–D7 arc connected to the outputs of Buffer1, Buffer2, and Buffer3. These data bits also drive the inputs of Latch1 and Latch2. The buffers are enabled only when the $\overline{OE}$ signal is taken low. Otherwise, the outputs of the buffers are in a tristate condition. The latches operate when data is presented on the inputs and the clock input is pulsed low. The rising edge transfers the data on the inputs to the output pins. Besides the output signals from the printer adapter, the block diagram also shows that one of the STATUS port bits can be used to generate an interrupt under control from one of the CONTROL port bits.

Table 3.2   The signals of the Centronics parallel printer adapter.

DB-25	Centronic	Register	I/O	Bit	Name	Function
1	1	Control	Out	C0*	nSTROBE	Active Low. Indicates valid data is on the data lines
2–9	2–9	Data	Out	D1–D8	DATA_1–DATA_8	Eight data lines. Output only in older SPP
10	10	Status	In	S6	nACK	A low asserted pulse to indicate that the last character was received
11	11	Status	In	S7*	BUSY	A high signal asserted by the printer to indicate that it is busy and cannot take data
12	12	Status	In	S5	PE	Paper empty
13	13	Status	In	S4	SELECT	Asserted high to indicate that the printer is online
14	14	Control	Out	C1*	AUTO FEED	Active low. Instructs the printer to automatically insert a line feed for each carriage return

Table 3.2 shows the signals of the actual parallel printer adapter (along with some other details). Table 3.3 shows some of the commonly encountered addresses for the three ports. However, one need not guess the port address for a particular system. It is possible to find out the exact port address from the system information in a PC.

The DATA Port

Figure 3.1 shows that the DATA port section of the adapter consists of Buffer1 and Latch1. When the CPU wants to transmit data to the printer, it write eight bits into the DATA port latch. The latch outputs are labeled D0–D7. D0 is the least significant bit, and D7 is the most significant bit. The latch outputs are available on pins 2–7 of the output connector. The DATA port output signals are also connected to the input of Buffer1. Buffer1 works as a read-back input port (see example in the Appendix). The

Table 3.2 (continued)

DB-25	Centronic	Register	I/O	Bit	Name	Function
15	32	Status	In	S3	nERROR	Signal by printer to the computer to indicate an error condition
16	31	Control	Out	C2	nINIT	Active low. Used to reset printer
17	36	Control	Out	C3*	nSELECT- IN	Active low. Used to indicate to the printer that it is selected
18–25	19, 21, 23, 25, 27, 29, 30, 34				GROUND	

Table 3.3 Typical port addresses for the three ports of the Centronics printer adapter.

Port	LPT1: PC	LPT2: PC	LPT3: AT
DATA	3BChx	379h	278h
STATUS	3BDh	379h	279h
CONTROL	3BEh	37Ah	27Ah

IBM PC/AT technical manual refers to this buffer as the data wrap buffer. The hex address of this port is x78h or x7Ch. (The x could be 2 or 3.) Thus, writing a byte to this address causes a byte to be latched in the data latch, and reading from this address sends the byte in the data latch to the microprocessor.

Figure 3.2 shows the output details for the DATA port. (Figure 3.2 is adapted from the IBM PC/AT technical manual.) Data is written to a device labeled 74LS374, which, according to the data sheets, is a tristate, octal, D-type, edge-triggered flip-flop (referred to as the latch). On the positive (rising edge) transition of the clock signal, the outputs of this device are set up to the D inputs. The outputs of the IC are the DATA port outputs available on pins 2–7 of the D-type output connector. The clock input of the latch is connected to the data write output of the command decoder block. This signal is activated anytime the microprocessor executes a port write bus cycle with the port address x78h or x7Ch.

The outputs of the latch are filtered through a 27-ohm resistor and a 2.2-nF capacitor before connecting to the output connector pins. The RC circuit slows down the rising and falling edge of the output voltage of the DATA port. The RC circuit ensures that any voltage transition (a high to low as well as a low to high) is gentle and not abrupt.

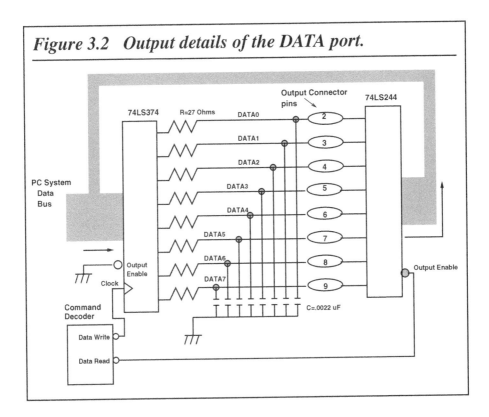

Figure 3.2   Output details of the DATA port.

An abrupt voltage transition on the printer cable is likely to induce noise on other DATA port lines or other signal lines and corrupt the data. With the RC circuit, this likelihood is reduced. The values of the resistor and the capacitor provide a time constant of about 60ns. With this time constant, very fast glitches (of durations less that 100ns) would be removed. The output drive capacity of the 74LS374 latch is as follows:

- Sink current: 24mA maximum,

- Source current: −2.6mA maximum,

- High-level output voltage: 2.4V DC minimum, and

- Low-level output voltage: 0.5V DC maximum.

The wrap-back buffer or the read-back buffer input pins are connected to the connector pins directly. This buffer is a 74LS244 IC, which is an octal tristate line receiver. When a data read instruction is executed by the microprocessor, the output enable pin of the IC is enabled and the data on the D-type connector pins is read into the microprocessor.

For simple parallel ports, the wrap-back buffer can be justified only as a diagnostic tool. Because the DATA port pins drive a cable, it is likely that the DATA signals could get shorted to ground or +5V (inside the printer or the destination circuit). The logic on the pins could then be permanently at 0 or 1. With the wrap-back buffer, the software can detect if the pins are stuck on some logic. On bidirectional and other advanced parallel ports (discussed in the next chapter), the wrap-back buffer is not merely for diagnostics. In these advanced designs, the wrap-back buffer reads external data, not necessarily data generated by the adapter.

Some programmers have used the ordinary DATA port bits for data input. The idea is as follows: At the start, the DATA port output latch is written with 0xffh (255 decimal). The pins of the port are then connected to the external device. As long as the output device can override the voltage set by the latch output, the data wrap-back buffer will read the data generated by the external device. However, this situation puts a lot of stress on the 74LS374 latch. I do not recommend this trick for data input on the DATA port.

The CONTROL Port

The CONTROL port of the adapter provides the necessary control signals to the printer. As shown in Table 3.2, the CONTROL port has four outputs on the output D-type connector. The CONTROL port has another signal that is not available on the connector, and that is the IRQ EN signal, which the driver program can use to enable interrupt generation with the help of the STATUS port signal (nACK), as described in the next section.

Figure 3.3 shows a block diagram of the CONTROL port. This block diagram is adapted from the IBM PC/AT technical manual. The output drivers of the CONTROL port signals are open collector inverters (referred to as OCI in the diagram). The open collector inverters are pulled up with resistors of the value 4.7Kohm. One of the outputs of the CONTROL port, C0* (nSTROBE), is also filtered using an RC circuit similar to the ones used at the DATA port output.

The output of the CONTROL port is derived from a hex D flip-flop IC, 74LS174. The data for the CONTROL port is latched by the low-going pulse from the command decoder, Control Write, to the 74LS174 IC. Three of the outputs of the IC, C0, C1, and C3 are inverted by an open collector inverter (labeled OCI in the diagram) IC.

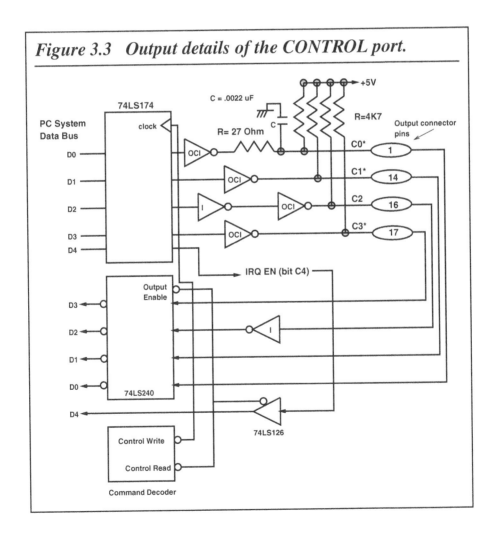

Figure 3.3 Output details of the CONTROL port.

These open collector drivers are pulled up with 4.7Kohm resistors. The other output of the CONTROL port, C2, is inverted by an ordinary inverter (labeled I in the diagram) before driving an OCI driver. The fifth output of the CONTROL port is the IRQ EN signal, which is not available on the output D-type connector but is used to enable or disable the interrupt generation from one of the STATUS port signals (nACK) as described below,

The state of all the CONTROL port signals can be read back using the wrap-back buffer IC, 74LS240, as shown in the diagram. This IC inverts the signals that pass through. The IRQ EN signal is read using a 74LS126 tristate buffer. The read process is controlled by the control read signal from the command decoder. The way the circuit is set up, the wrap-back buffer provides the same state of the CONTROL signals as have been latched in the 74LS174 IC.

Appropriate inverters are included to cancel any signal inversion in the read-back path. Thus, if the microprocessor sends the following byte to the CONTROL port latch:

Data bits:	D7	D6	D5	D4	D3	D2	D1	D0
Value:	0	1	0	0	0	1	1	0

the wrap-back register provides the following byte:

Data bits:	D7	D6	D5	D4	D3	D2	D1	D0
Value:	X	X	X	0	0	1	1	0

The X in the last three bits means that these bits have an indeterminate state owing to the fact that only five bits are asserted by the wrap-back register when the control read operation is performed by the microprocessor.

It is interesting to note that the output drivers of all the CONTROL port signals are open collector devices with pull-up resistors. The advantage of such a configuration is that it allows the CONTROL port bits to read external data. To use the CONTROL port as an input port, the microprocessor sets the CONTROL port outputs to logic 1s. If any external logic is applied to these pins, the wrap-back input port for the CONTROL port will read the logic state applied by the external source.

The STATUS Port

The STATUS port provides the printer adapter with the facility to read the status of the printer through various signals. The STATUS port is at an address next to the DATA port. Typical hex addresses are 0x379h or 0x3BDh. The STATUS port signals are labeled S7 for the most significant bit to S0, though S0 does not exist. Signals up to S3 are available. The STATUS port signals have the following functions.

S7* (BUSY)   This signal from the printer indicates that the printer is busy and cannot take more data. It is important to note that this signal is inverted by a NOT gate on the adapter board. That is why the signal is labeled S7* and not S7. The implication of this inversion is that low voltage applied on the connector pin will be read as a high voltage by the microprocessor.

S6 (nACK)   This is a signal that the printer generates in response to the strobe signal from the adapter as an acknowledgment. Normally the signal is high and after the printer is strobed, the printer responds by taking this signal low and then high again.

S5 (PE)   This is a signal that the printer generates to indicate that there is no paper in the printer. Normally the signal is held low by the printer, and when the printer paper is exhausted, the signal goes high.

S4 (SELECT)   This signal is asserted high by the printer to indicate that the printer is enabled. When the printer is disabled (to feed or advance paper or to change a font on the printer), this signal is low.

S3 (nERROR)   This is a general error indicator signal on the printer. There could be many reasons for the cause of the error, such as jammed paper or an internal error. In the event of an error, the signal is set to low voltage.

The STATUS port signal S6 (nACK) can be routed with the help of a signal from the CONTROL port, such that it acts as an interrupt input (IRQ7 or IRQ5). The rising edge of the nACK signal (a low to high transition) will generate, if enabled with the appropriate CONTROL port bit, a low to high transition on the IRQ7 or the IRQ5 interrupt line. If the particular IRQ (5 or 7) has been enabled by the program, the CPU will execute an Interrupt Subroutine.

The original designers of the printer adapter probably thought of allowing an interrupt-driven printer driver. After each character is transferred to the printer, the printer acknowledges by generating a high-to-low-to-high pulse on the nACK line and the low-to-high transition could trigger an interrupt that would transmit the next character to the printer. However, given the latency in executing the interrupt subroutine after the interrupt is generated — on the original PC, the worst case interrupt latency is more than 110µs— it was not practical to use an interrupt-based printer driver. Generally, IRQ5 and IRQ7 can be used for other applications.

Figure 3.4 shows the output details of the STATUS port. This block diagram is adapted from the IBM PC/AT technical manual. The figure shows that four of the STATUS port bits are connected to a 74LS240 IC, which is an octal tristate buffer with output inverters (as shown by the bubbles at the output of the IC). Three of the four signals to this IC are passed through a NOT gate. These inverters cancel the further inversion that the signals encounter at the output of the LS240 buffer IC. Only the BUSY signal goes to the buffer IC uninverted. The logic of the BUSY signal is inverted when it is received by the microprocessor. The fifth signal, nERROR, is received by a tristate buffer 74LS126. This signal is also transmitted to the PC system data bus in an uninverted state. Thus, of all the STATUS port signals, only the BUSY signal is inverted and so is referred to as S7*. The other signals are S6, S5, S4, and S3. As can be seen from the figure, the S6 (nACK) signal of the STATUS port can also be used to generate an interrupt on the IRQ5 or IRQ7 line under control of the CONTROL port bit 4 (IRQ EN). If the IRQ EN bit of the CONTROL port is high, it enables the 74LS126 buffer, and the nACK signal passes to the IRQ5 (or the IRQ7) input pin of the Programmable Interrupt Controller IC (8259).

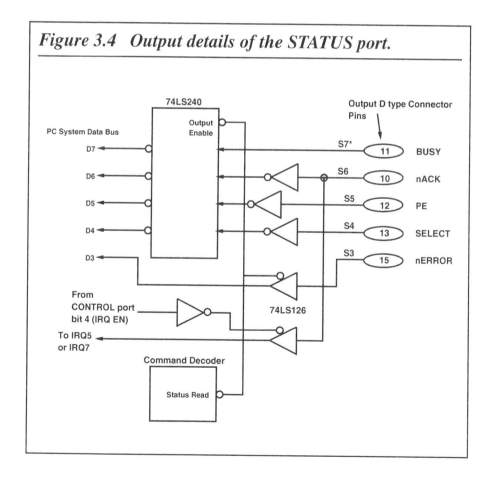

Figure 3.4   Output details of the STATUS port.

Printing with the Parallel Adapter

Figure 3.5, a timing diagram for the parallel printer adapter, shows how only three signals, besides the DATA port signals, can be used to transmit data to the printer and to read back the status of the printer. The signals are: nSTROBE, BUSY, and nACK.

Unlike a microprocessor system bus, the DATA port signals are always active (the data bus signals on typical microprocessor systems can go in tristate). The PC sets the DATA port signal lines with the appropriate logic and after some delay, typically 0.5μs, generates the nSTROBE signal. The nSTROBE signal is a low-going pulse, typically of 0.5μs duration. After that, the nSTROBE signal is pulled high. The data on the DATA port signal lines is held valid for some time even after the nSTROBE signal is pulled to logic high. Again, the data on the DATA port signals is held valid for about 0.5μs. This ensures that data is properly latched in any device with a requirement of nonzero data hold time.

In response to the low-going nSTROBE pulse from the PC, the printer responds by setting its BUSY signal high. The printer can hold the BUSY signal high for an indeterminate amount of time. After the data is evaluated and used by the printer, the printer pulses the nACK signal to say that it is ready for more data.

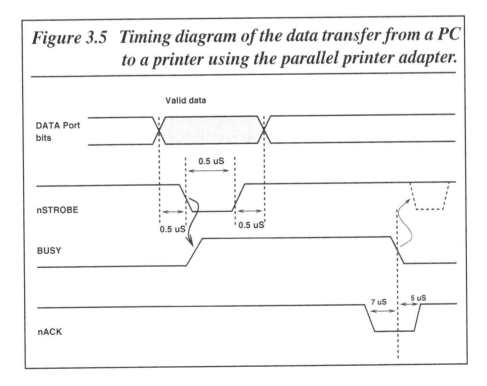

Figure 3.5 Timing diagram of the data transfer from a PC to a printer using the parallel printer adapter.

Typically, about 7µs after the low-going nACK pulse, the printer takes the BUSY signal low again. The PC transmits more data only when it detects that the BUSY signal is low. After a delay of 5µs from the time the BUSY signal is taken low, the printer puts the nACK signal high again.

Using the Parallel Printer Adapter

With a standard parallel port, you have access to a 5-bit input port, a 4-bit output port, and another 8-bit output port. The 4-bit output port can also be configured as a 4-bit input port. All 17 signal lines are accessible under program control and can be used for TTL signal level data transfer in to and out of the PC. An interrupt input signal is also available. The parallel port signals are available on the rear panel of PCs on a 25-pin D-type female connector. To connect a PC and a printer, a ready-made cable with a 25-pin D-type male connector on one end and a 36-pin Centronics male connector on the other end is available. This cable can also be used to connect the parallel port to any other circuit.

However, it is important to choose good-quality cables. Many of the cheap cables are reported to have a few signal lines missing. This may go undetected if you only need to connect the PC and a printer, but for any other use, missing signal lines could be disastrous. You may want to make your own cable of the required length and quality. (The 1284 standard does define a compliant, high quality cable, which should be used.)

Chapter 4

Programming and Using the Parallel Adapter

This chapter looks at problems associated with programming and using the parallel port. The C programs in this chapter and in later chapters have been tested (unless otherwise indicated) with the Turbo C v2.0 compiler. I believe these programs can be used without any change with the higher versions of the Turbo C or Borland C compilers and, after minor modifications, with other compilers such as the Microsoft C compiler. The later sections of this chapter provide examples for connecting hardware to the parallel port.

PC Data Area

When you decide to use the parallel port for a nonprinting application, the first thing you must determine is the address of the parallel port adapter. Most PCs use one of three possible parallel port addresses. Fortunately, MS-DOS stores the addresses of most hardware in a specific area of memory called the BIOS data area. The base addresses of the parallel adapters are also available in this area. The PC recognizes up to three logical parallel adapters. By examining this area, the user can find out if the PC is equipped with a parallel adapter and, if so, the base address of the parallel adapter.

Listing 4.1 is a routine that reads the memory of the PC and finds out the base address of the parallel adapter. The program allows the user to detect whether the PC is equipped with a parallel adapter and, if so, requests the user to enter a number 1, 2, or 3. Accordingly, the program reads the BIOS area and determines if the parallel adapter is present. If the parallel adapter is present, the value read is nonzero and is the base address of the adapter. The base address of the adapter is the port address of the DATA port. The STATUS port is at an address location one higher than the DATA port address, and the STATUS port is at an address location two higher than the DATA port address. Listing 4.1 is a stand-alone piece of code that can be executed on a DOS-based PC to detect if a printer adapter is present. In real application circuits, this piece of code is vital for the functioning of the add-on circuits.

Accessing the Ports

It is important to realize that two of the parallel port's internal ports, the STATUS and CONTROL ports, are incomplete (i.e., of a possible eight bits, these ports have only five and four signals, respectively, available on the output connector). Also, many of the port bits have an inverter between the bit and the output connector pin. The DATA port is free of any such intrigue and can be used very cleanly.

To transfer data to an output port, use the function macro

```
outportb(port_address, data);
```

The first argument of the function macro is the address of the destination port, and the second argument is the data to be transmitted. The first argument can be a constant or a variable of the type int and the second argument is a constant (a byte of data) or a variable of type char.

To read data from an input port, use the function macro

```
data_variable = inportb(port_address);
```

The instruction returns a byte that is transferred into the variable data_variable, which should be of type char.

A Break-Out Box for the Parallel Adapter: Lighting LEDs and Reading Switches

A break-out box is a piece of equipment that is usually used to simplify troubleshooting. Break-out boxes for RS-232 serial ports are very common. The break-out box I will describe is a simple circuit with a few LEDs, resistors, and a couple of jumpers, mounted on a small PCB with a male 25-pin D-type connector, that can be plugged into the PC's parallel adapter connector. The purpose of this break-out box is to acquaint you with the working of the parallel adapter ports (Listing 4.2), in particular of the STATUS and CONTROL ports.

Listing 4.1   Find the base address of the parallel adapter.

```c
/*detect.c*/
/*This piece of code determines if the PC has any parallel adapter*/
/*If yes, it finds out and prints the addresses of the DATA, CONTROL */
/*and the STATUS ports*/

#include <stdio.h>
#include <dos.h>

main()
{
    int dport, cport, sport, select, offset;
    clrscr(); /*clear screen*/
    printf("\tProgram to detect the parallel printer adapter.\n");
    printf("Enter 1, 2 or 3 to detect LPT1, LPT2 or LPT3 respectively.\n\n");

    /*wait for a keystroke*/
    select=getch();

    /*loop indefinitely if the key pressed is not '1', '2' or '3'*/
    while ( (select != '1') && (select != '2') && (select != '3') )
        {
        printf("Invalid number. Enter 1, 2 or 3\n");
        select=getch();
        }
    if(select == '1') offset = 0x08;
    if (select == '2') offset = 0x0a;
    if (select == '3') offset = 0x0c;

    /*now look into BIOS area to determine the address of the particular*/
    /*parallel adapter*/
    dport=peek(0x40, offset);

    /*If the address is zero, means that particular parallel adapter*/
    /*is not present. No point in continuing. Abort the code here*/
    if (dport == 0)
        {
        printf("Sorry, On this machine LPT%d does not exist\n", select-'0');
        exit(0);
        }

    /*else print the port addresses of the DATA, STATUS and CONTROL ports*/
    printf("\nLPT%d detected!\n", select -'0');
    printf("DATA port is %x\n", dport);
    printf("STATUS port is %x\n", dport+1);
    printf("CONTROL port is %x\n", dport+2);
}
```

As you can see from the circuit diagram in Figure 4.1, the LEDs are connected to the outputs of the DATA port as well as to the outputs of the CONTROL port. The outputs of the CONTROL port are connected to the LEDs through a set of two-way jumpers that allow the CONTROL port signals to either connect to the LEDS or to the signals of the DATA port. The STATUS port signals are connected to the DATA port signals through another set of jumpers. With this jumper switch arrangement, the STATUS port signals can be disconnected from the DATA port signals. This allows you to program the DATA port to an arbitrary sequence and to read this sequence back through the STATUS port.

Figure 4.1   Circuit diagram for the parallel adapter break-out box.

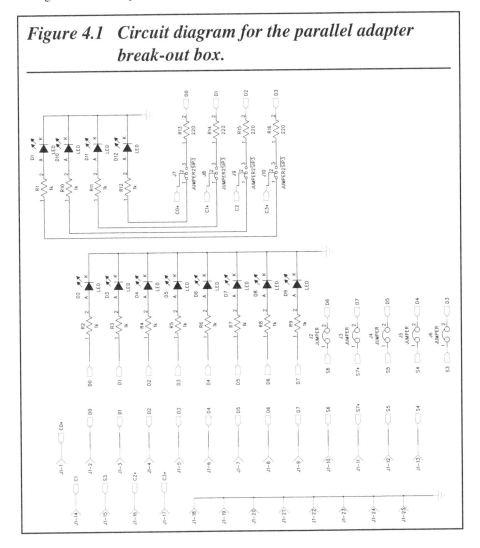

Listing 4.2   *Play with the parallel ports using the break-out box.*

```
/*bob.c*/
#include <dos.h>
#include <conio.h>
#include <stdio.h>

unsigned char selection;

void play_dport(void)
{
   static unsigned char a;
   /*repeat till a key is pressed*/
   while(!kbhit())
     {
     outportb(0x378, a);
     printf("\nDATA port = %X",a);
     a++;
     sleep(1);
     }
  getch();
}

void play_cport(void)
{
  static unsigned char c;
  while(!kbhit())
  {
    outportb(0x37a, c);
    printf("\nCONTROL port = %X",c);
    c++;
    sleep(1);
  }
  getch();
}

void dout_cin(void)
{
  static unsigned char dout, cin;
  outportb(0x37a, 0x04);
  while(!kbhit())
  {
    outportb(0x378, dout);
    cin=inportb(0x37a) & 0x0f;
    printf("\nDATA port = %X, CONTROL port raw = %X,
           CONTROL port corrected = %X", dout & 0x0f, cin, 0x0b ^ cin);
    dout++;
    sleep(1);
  }
  getch();
}
```

Listing 4.2 (continued)

```
void dout_sin(void)
{
  static unsigned char dout, sin;
  while(!kbhit())
  {
    outportb(0x378, dout);
    sin=inportb(0x379) & 0xf8;
    printf("\nDATA port = %X, STATUS port = %X,
           STATUS port corrected = %X", (dout & 0xf8)>>3, sin>>3,
           0x80 & (sin>>3));dout=dout+8;
    sleep(1);
  }
  getch();
}

unsigned char get_selection(void)
{
        puts( "\nPlay with DATA Port\t\t\tD\n" );
        puts( "DATA port output, CONTROL port input\tC\n" );
        puts( "DATA port output, STATUS port input\tS\n" );
        puts( "CONTROL port output\t\t\tO\n" );
        puts( "Exit \t\t\t\tX\n" );
        printf( "Enter selection: " );
        gets( &selection );
        selection |= 0x20;
        return selection;
}

main()
{
  for(;;){
  get_selection();
        switch (selection) {
            case 'd': printf("\nCase D\n");
                    play_dport();
                    break;            case 'c': printf("\nCase C\n");
                    dout_cin();                    break;
            case 's': printf("\nCase S\n");
                    dout_sin();
                    break;
            case 'o': printf("\nCase O\n");
                    play_cport();
                    break;
            case 'x': printf("\nBye... Play me again!\n");
                  exit(0);
            default: printf("\nError\n");
        }
    }
  }
```

The DATA port signals are also connected to the CONTROL port signals to show how the CONTROL port can be used to read external data. With the two-way jumper switches connected to the CONTROL port signals, these signals can either light the LEDs in one position of the switch or connect to the DATA port signals in the other switch position. DATA port signals are connected to the CONTROL port through 220-ohm resistors, This is to ensure that if the CONTROL port output is 0, the DATA port signal outputs are not pulled low.

Power Switching Circuits for the Parallel Adapter

To control real-world devices, you need to control power through the devices either in an on/off mode or in a continuously (linearly) varying mode. Simple devices can be controlled through relays and solenoids.

To control relays (electromechanical as well as solid-state) and solenoids, you need some sort of switch, a transistor switch for example. The outputs of the parallel ports use transistor switches. However, these switches cannot handle voltages and currents more than 5V and a few milliamps — most relays need substantially larger voltages and currents to operate.

To activate relays, you need switches with high current and voltage capacity. There are many types of switches that can be configured to meet a specific require-ment. Rather than go in for switches made out of discrete components, it is worth-while to hunt for fully integrated switches. Sometimes, however, it may be the case that your requirements cannot be met with integrated components and you may be forced to look for discrete solutions.

You may encounter a need to switch currents at very high frequencies, as is the case of switches used to drive stepper motors. With advances in MOSFET technology, high-speed and high-power switches using MOSFETs have become more common and popular compared to transistor switches. The problem with MOSFET switches is the need for suitable drivers. Because MOSFETs turn on and off, voltages are differ-ent from those available from digital outputs, so driving MOSFETS can sometimes be tricky. However, TTL-compatible MOSFET drivers can alleviate such problems.

ULN2003A Driver

The ULN2003A are high-voltage, high-current darlington arrays containing seven open collector darlington pairs with common emitters. Each of the seven channels can handle 500mA of sustained current with peaks of 600mA. Each of the channels has a suppression diode that can be used while driving inductive loads (such as relays) as freewheeling diodes.

The ULN2003A input is TTL compatible. Typical uses of these drivers include driving solenoids, relays, DC motors, LED displays, thermal print heads, etc. The IC is available in a 16-pin DIP package and other packages. The outputs of the drivers can also be paralleled for higher currents, though this may require a suitable load-sharing mechanism.

Figure 4.2 shows the block diagram of the ULN2003A darlington array driver IC. For each of the drivers, there is a diode with the anode connected to the output and the cathode connected to a common point for all seven diodes. The outputs are open-collector, which means that external load is connected between the power supply and the output of the driver. The power supply can be any positive voltage less than +50V, as specified by the data sheets. The load value should be such that it needs sustained currents less than 500mA and peak currents less than 600mA/driver.

Figure 4.3 shows three relays being driven by the outputs of three drivers from the ULN2003A IC. One end of the relay coil is connected to the output of the driver and the other end is connected to the +ve supply voltage. The value of this voltage will

Figure 4.2   ULN2003A darlington array.

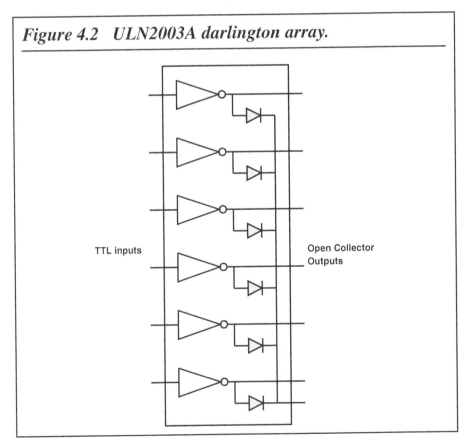

TTL inputs

Open Collector
Outputs

depend upon the relay coil voltage ratings. The diode common point is also connected to the +ve supply voltage. The input to the ULN2003A IC is TTL voltages — the output of the DATA port of the parallel adapter, for example. With this arrangement, the DATA port signals could be used to control each of the relays.

The relay terminals labeled NC (Normally Closed), common, and NO (Normally Open) could be used to switch whatever voltage may need to be switched. Typically, the relay terminals are used to switch the mains supply (220V AC or 115V AC, as may be the case) to the required load (a heater, a lamp, etc.), but of course, the relay may be used to switch any voltage (AC or DC) as long as the relay contact can handle the voltage and the current.

MOSFET Drivers

Using MOSFETs for power switching is getting extremely popular and there are reasons for that. It is relatively easier to use MOSFETs than bipolar transistors (BJT) for large switching currents. Power MOSFETs are getting cheaper than the bipolar transistors for

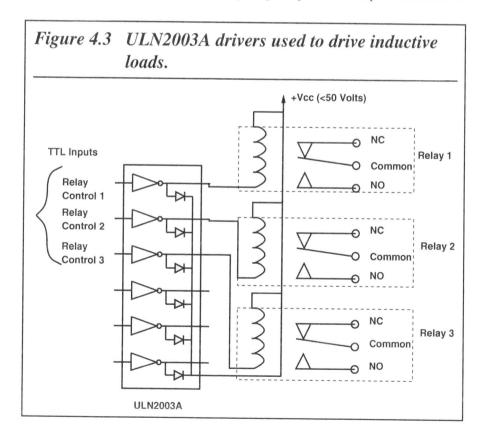

Figure 4.3 ULN2003A drivers used to drive inductive loads.

comparable voltage and current characteristics. However, it is very easy to encounter some pitfalls while using MOSFETs. This chapter examines some of the problems and possible solutions.

MOSFETs have become popular for four reasons:

Speed   MOSFETs are inherently faster than BJTs because only a single majority carrier (electrons or holes) is involved. The switching speeds of the MOSFETs depend upon the charging and discharging of device capacitances (which are not insignificant), independent of temperature.

Input characteristics   The MOSFET input, the gate, is electrically isolated from the device source and has a DC resistance of tens of megaohms. The MOSFETs are typically turned on at 10V or less (there are some varieties called logic level switching types which turn on at 2V and are suitable to be driven directly off the output of a TTL gate). The gate drive requirement is nearly independent of the load circuit.

Operating parameters   Unlike BJTs, MOSFETs do not require derating of the power handling capacity as a function of applied voltage. The operating levels are clearly determined by the peak current ratings, power capabilities, and the breakdown voltages of the devices.

On voltage   The On voltage is determined by the device On resistance (resistance between the drain and source of the MOSFET). For low power MOSFETs, this is small and for power MOSFETs this can be somewhat large. But a large On resistance has the benefit that many devices can be operated in parallel with load sharing.

MOSFETs work like BJTs. To turn an NPN BJT on, you need to supply a base current and a base-emitter voltage that is greater than the threshold voltage (between 0.7 and 1.0V). The base current required to turn the BJT on depends upon the load current and collector current to base current ratio (called the ß). Similarly, for N-channel MOSFETs, you need to apply a gate voltage greater than the gate source threshold. At DC, the required gate current is very small (on the order of microamps) and is not a problem, but at higher frequencies, the input capacitance of the MOSFET (which is much larger than that of the BJT) requires transient current, which can be much larger than that of a comparable BJT. The popular belief that MOSFETs only require a very small, negligible gate current is shattered once you look at the following figures. Take a typical case of the input capacitance of a big power MOSFET as 5,000pF. Assume that the gate voltage must switch from 0 to 10V in 1µs (so that the output load can switch in that time). To charge a capacitance of 5,000 pf to 10V in 1µs requires a charging current, I_g, which is

$$I_g = C \bullet V/t$$

which is

$$I_g = 5000 \bullet 10^{-12} \bullet 10/10^{-6} = 50\text{mA}$$

This is not an insignificant amount of current and definitely not something a typical logic gate can provide, so to drive MOSFETs with logic outputs, you need to pay attention to the turn-on current requirement at the required switching rate and the turn-on threshold voltage. If these cannot be provided by the output of the logic device, an intermediate driver must be used.

Thus, driving MOSFETs from digital logic outputs not only require a sufficient voltage to turn the MOSFET on but may also require significant amounts of gate currents. As far as turn-on gate voltage, a 10V gate drive will turn almost any MOSFET on. Many MOSFETS have much lower (1.0–1.5V) gate threshold voltages and these are specially suitable for use with TTL digital outputs.

Figure 4.4 MOSFET symbols and a MOSFET-controlled mechanical shutter.

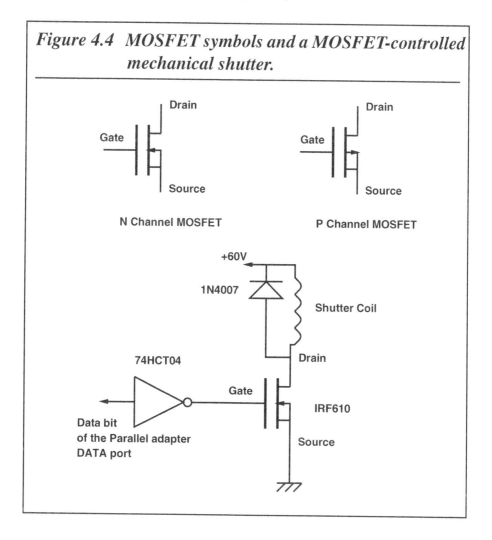

Figure 4.4 shows the symbols of N- and P-channel MOSFETS. Figure 4.4 also shows a simple circuit to drive a mechanical shutter coil that operates at +60V power supply. The N-channel MOSFET IRF610 which has low Vgs voltage, is used to drive the shutter coil. The gate of the MOSFET is driven by the output of a 74HCT04 inverter. The inverter is driven by one of the bits of the parallel adapter DATA port bits.

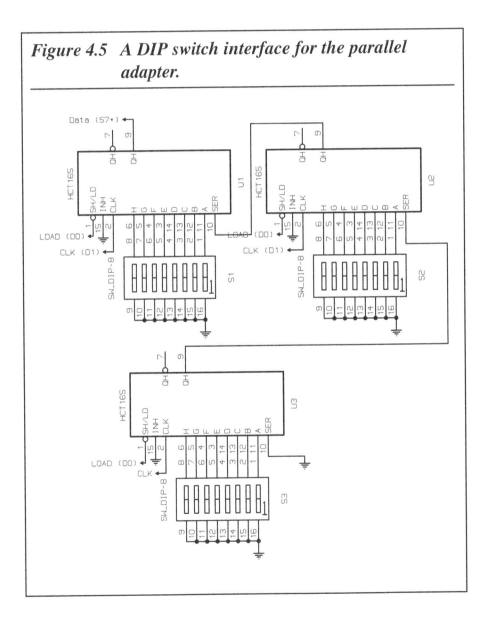

Figure 4.5   A DIP switch interface for the parallel adapter.

Reading DIP Switches

There are times when you need to read DIP switches. DIP switches are useful input devices very commonly used to implement user choice. Figure 4.5 shows the circuit to interface three DIP switches of eight bits each to the parallel adapter. Only three signal lines of the parallel adapter are used.

The heart of the circuit is a 74HCT165 parallel-in, serial-out shift register. Many of these shift registers can be cascaded using the SER signal pin. The shift register has an input latch that is loaded with the logic levels presented at the input pins (A–H) by application of the LOAD signal. In this circuit, one terminal of the DIP switches is connected to the input of the shift register, and the other terminal is grounded. When the switch is open, the input to the shift register is high (for TTL-compatible logic only, for other logic family, use pull up resistors of 1Kohm value). When the switch is closed, the logic input is low.

The other input to the shift registers is the CLK input. Each edge of the CLK signal shifts the data stored in the latch to the output QH pin of the IC. The output of the shift register is connected to the STATUS port bit S7*.

The DIP switch reader program begins by issuing the LOAD pulse and then generates 24 CLK signals. After each CLK signal it reads the status of the S7* pin and stores it in a variable. In the end, all the bits are separated and put in relevant variables. Listing 4.3 is the driver program for reading three DIP switches of eight bits each.

Listing 4.3 Driver program to read DIP switches.

```
/*rd_dip.c*/
#include <stdio.h>
#include <dos.h>
#include <conio.h>
#include <process.h>

/* This programme allows the computer to read a set of DIP switches
   These switches are connected to corresponding shift registers,
   which convert their parallel data into serial .
   The serial data is read at the printer (parallel) port of the PC
   The MSB of the printer Status port (S7*) is used as the input data bit */

                                    /* D3  D2 D1  D0  */
#define RESET_VALUE    OXFF         /* 1   1  1   1   */
#define GEN_LOAD       OXFE         /* 1   1  1   0   */
#define GEN_CLOCK      OXFD         /* 1   1  0   1   */
```

Listing 4.3 (continued)

```c
#define DIP_SW 3        /*total number of DIP switches*/
#define BITS_SW 8       /*bits in 1 DIP switch */

void main(void)
{
    int dport_lpt1, sport_lpt1, shift_reg[6], temp, in_count, out_count,
        columns, rows;

    /* Sign ON */
    clrscr();
    printf("DIP Switch reading program.");
    printf("\nReads the Shift register connected to Printer Control & Status Port");
    printf("\nLOAD ---> D0.");
    printf("\nSHIFT --> D1.");
    printf("\nDATA ---> S7.");
    printf("\n\n\nD.V.GADRE");

    /*Get LPT1 port addresses */
    dport_lpt1 = peek(0x40,0x08);
    if(dport_lpt1 ==0)
      {
      printf("\n\n\nLPT! not available... aborting\n\n\n");
      exit(1);
      }
    printf("\n\n\nLPT1 address = %X\n\n", dport_lpt1);
    sport_lpt1 = dport_lpt1 + 1;    /* status port address */

    /* Initialize the Printer DATA Port*/
    outportb(dport_lpt1, RESET_VALUE);

    /* kbhit to see of keyboard is hit */
    /* come out of loop, if so*/
    while (!kbhit())
      {

      /* Generate LOAD signal on D0 bit for the shift registers */
      outportb(dport_lpt1, GEN_LOAD);
      outportb(dport_lpt1, RESET_VALUE);

      for(out_count=0; out_count<DIP_SW; out_count++)
          /* number of DIP Switches */
          {

          for(in_count=0; in_count<BITS_SW; in_count++)
              /* number of bits / Switch */
              {
```

Data Transfer Overheads Using the Standard Parallel Port

The previous chapter described how the standard parallel port is used to transfer data to a printer. This section will examine the problem of data transfer from a software perspective. This exercise, which will help explain why data transfer using the standard printer port is slow and why vendors and developers felt the need to develop a faster interface, will lead to the reasons behind the development of the Enhanced Parallel Port (EPP) and the Extended Capability Port (ECP) — topics for the next chapter.

Listing 4.3 (continued)

```
            /* Read the Status Port bit S7*/
            temp = inportb(sport_lpt1) ;
            temp = temp ^ 0x80;          /*invert bit no. 7*/
            temp = temp >> 7;    /*shift it to the lower most position*/
            temp = temp & 0x01;  /*mask all bits except the last bit*/

            /* Concatenate it in the variable */
            shift_reg[out_count] = shift_reg[out_count] << 1;
                /* shift the bit one left*/
            shift_reg[out_count] = shift_reg[out_count] & 0xfffe;
                /* make the LSB 0*/
            shift_reg[out_count] = shift_reg[out_count] | temp;
                /* place the new bit in the LSB*/
            shift_reg[out_count]=shift_reg[out_count] & 0xff;
                /* keep only the lowest 8 bits*/
            outportb(dport_lpt1, GEN_CLOCK); /* clock for the next bit*/
            outportb(dport_lpt1, RESET_VALUE);
            }
        printf("Switch %d = %X\n", out_count, shift_reg[out_count]);
        }
    printf("\n");
    delay(500); /*wait for .5 seconds*/
    }
    /* Repeat loop */
    return;
}
```

Listing 4.4   Transfer data to an external device.

```c
/*prin_dat.c*/
#include <stdio.h>
#include <dos.h>
#include <conio.h>
#include <process.h>

#define BUF_LEN 1000

/*Global variables that store the addresses of the three ports of the
 standard printer adapter*/

unsigned int dport, cport, sport;

main()
{
    /*the following array stores data to be transfered to the external device*/
    unsigned char print_buffer[1000], temp;
    unsigned int count;

    /*Get LPT1 port addresses */
    dport = peek(0x40,0x08);

    if(dport == 0)
        {
        printf("\n\n\nLPT! not available... aborting\n\n\n");
        exit(1);
        }
    printf("\nLPT1 address = %X", dport);
    cport = dport +2;     /* control port address */
    sport = dport + 1;    /* status port address */

    /*this statement puts all the CONTROL port signals to logic 1*/
    outportb(cport, 0x04);

    /*setup a loop to transfer the required data points*/
    for(count=0; count<BUF_LEN; count++)
        {
    /*First check if the external device is not BUSY*/
    /*the BUSY signal is the msb of the status port, so read the status
    port, isolate the 7th bit and wait till it becomes 1. Since the bit
    is inverted, a BUSY low means that the register bit is 1. In the
    following example some sort of timeout feature must be included for
    practical use*/
        temp=inportb(sport);
        temp=temp & 0x80;
```

Listing 4.4 gives a reasonable estimate of what it takes, from a software perspective, to transfer data from a PC to an external device like a printer. In Listing 4.4, you can easily see that to transfer even a single byte, the processor has to execute at least 10 instructions. Of these many instructions, at least five read and write to the three ports of the parallel printer adapter. Thus, on even the fastest PC, the bottleneck in data transfer rate is the port I/O instructions. As a result, data transfer rates using the standard printer adapter to an external device never exceed 200–300Kbps. If the PC needs to read byte data from an external device, consider your speed to be in the range of 100–250Kbps, at best. This is because the standard printer adapter can only read four bits at a time. So to read byte-wide data, it must perform two read operations, one to read the lower nibble and the other to read the higher nibble, and then assemble the two nibbles to get the byte. That's not all. The byte has the seventh and third bits inverted, so a bit inversion operation must also be done on this assembled byte to get the real byte. The following example shows how the standard printer adapter is used to read external data using the STATUS port signals.

Figure 4.6 shows a block diagram to read external data through the STATUS port signals. This example shows only the rudiments of a practical circuit. In this example circuit, I have not shown the external source of data or any scheme of synchronization between the source and the PC.

Listing 4.4 (continued)

```
        while(temp != 0x80)
            {
            temp=inportb(sport);
            temp=temp & 0x7f;
            }

        /*OK, now the external device is ready so pump some data*/
        outportb(dport, print_buffer[count]);

        /*now pulse the nSTROBE signal low and then high again*/
        /*make sure that the other CONTROL signals are not disturbed*/
        temp=inportb(cport);
        temp=temp | 0x01;        /*this makes the external signal low*/
        /*nSTROBE occupies the lsb of the CONTROL port register*/

        outportb(cport, temp);
        temp=temp & 0xfe; /*make it high again*/
        outportb(cport, temp);
    }
}
```

The external source produces data bits D0–D7 (please do not confuse these with the outputs of the DATA port of the printer adapter). These eight data bits are fed to the inputs of two 74HCT244 tristate buffer ICs. Each IC has two independent sections. Each of the sections has an enable signal, as shown by the bubble at the lower right side of the IC block. The circuit is arranged in such a fashion that, at any instant of time, only one of the two sections is enabled and the other is disabled. This is achieved by using the inverter output to generate one of the enable signals. The uninverted signal controls the other enable signal. Signal C0* from the CONTROL port is used to enable the buffer blocks.

Each output of the buffer ICs is shorted to a corresponding signal from the other block. Thus, from the eight inputs to the buffer IC, you get four signals. To read a byte, the PC puts the C0* signal to logic 0, reads the STATUS port, and temporarily stores it. It then changes the C0* signal to logic 1, and then reads the STATUS port once more. Reading the STATUS port each time, you get a byte, from which you have to discard the lower nibble. Then using the two nibbles, the program generates one byte, as shown in Listing 4.5.

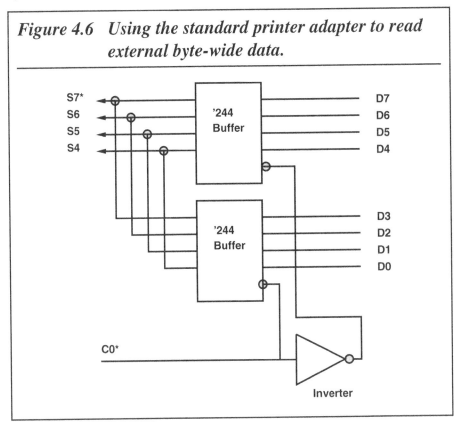

Figure 4.6 *Using the standard printer adapter to read external byte-wide data.*

The number of operations required to receive a byte of data through the STATUS port of a standard parallel adapter is even greater than the number of operations required to send data out through the DATA port. One way to address this problem is to have a parallel printer adapter with a bidirectional DATA port. IBM, for example introduced this feature in its PS/2 computers. Many of the clones today have this feature. Having a bidirectional DATA port saves many steps when reading data. However, it is still necessary to switch the direction of the DATA port to receive data. This would need to be done only once, provided no data needs to be sent out the DATA port during this time.

The next chapter looks at the need for high-speed parallel ports and the solution in terms of the EPP and the ECP.

Listing 4.5   Read external data using the STATUS port.

```c
/*read_dat.c*/
#include <stdio.h>
#include <dos.h>
#include <conio.h>
#include <process.h>

#define READ_LEN 1000 /*number of data points to read*/

/*Global variables that store the addresses of the three ports of the
standard printer adapter*/

unsigned int dport, cport, sport;

main()
{
    /*the following array stores data that is received from the external device*/
     unsigned char input_buffer[1000], low_nib, high_nib, real_byte, temp;

    unsigned int count;

    /*Get LPT1 port addresses */
    dport = peek(0x40,0x08);
    if(dport ==0)
         {
         printf("\n\n\nLPT! not available... aborting\n\n\n");
         exit(1);
         }
    printf("\nLPT1 address = %X", dport);
    cport = dport +2;     /* control port address */
    sport = dport + 1;     /* status port address */
```

Listing 4.5 (continued)

```
/*this statement puts all the CONTROL port signals to logic 1*/
outportb(cport, 0x04);

/*setup a loop to read the required data points*/
for(count=0; count<READ_LEN; count++)
    {
    /*in practice, some sort of synchronization code will go here*/

    temp=inportb(cport);
    temp=temp | 0x01;    /*this makes the external C0* signal low*/
    outportb(cport, temp);

    low_nib=inportb(sport);

    temp=temp & 0xfe;        /*make external C0* signal high again*/

    outportb(cport, temp);
    high_nib=inportb(sport);

    low_nib=low_nib >> 4;
    /*shift the 4 upper bits to lower 4 positions*/
    low_nib = low_nib & 0x0f; /*make sure upper 4 bits are all 0s */

    high_nib=high_nib & 0xf0;      /*to make lower 4 bits  all 0s */

    real_byte = high_nib | low_nib;   /* OR the two bytes together*/

    real_byte=real_byte ^ 0x88;
    /*flip the 7th and the 3rd bit and now the real byte is ready*/

    /*now store it*/
    input_buffer[count]=real_byte;
    }
}
```

Chapter 5

The Enhanced Parallel and Extended Cabability Ports

The parallel printer adapter is such a versatile port that people started using it for applications other than printing. The use of the parallel port soon began spreading to such applications as tape backup systems, CD-ROM players, and LAN adapters. However, the performance of the parallel printer adapter was severely limited by the data transfer rates from the PC to the peripheral (about 250Kbps maximum) and by the lack of a high-speed protocol to transfer data from the peripheral back to the PC. In addition, due to the lack of a standardized electrical interface, the maximum distance between the PC and the external peripheral was limited to about six feet.

Until about 1991, the features offered by the standard parallel port (in terms of speed, data buffering, etc.) didn't progress as quickly as the performance characteristics of host computers. In 1991 various printer manufacturers formed a group called the Network Printing Alliance (NPA). This organization sought to develop a standard for the control of printers across a network. As part of their work, NPA decided to look at the parallel port and to define new protocols that would increase performance and provide bidirectional communication while retaining backward compatibility with existing parallel ports.

The NPA wanted to implement a set of parameters that would require a high-performance bidirectional connection to the PC with backward compatibility, so they approached the IEEE with a request to constitute a committee with that task. The IEEE 1284 committee decided to aim for a bidirectional data transfer rate greater than

1Mbps. The committee released the standard "Standard Signaling Method for a Bi-directional Parallel Peripheral Interface for Personal Computers" in March 1994. This standard gave rise to the Enhanced Parallel Port (EPP).

This chapter discusses the EPP standard and provides details about the EPP mode of operation. I will describe how to transfer data in EPP mode with an example circuit and communication routines. I will also look at the Extended Capability Port (ECP) protocol and the proposed electrical interface for new parallel ports.

The IEEE 1284 1994 Standard

The IEEE 1284 standard (hence forth referred to as 1284) provides bidirectional communication between the PC and external peripheral devices at a rate of 20–50 times faster than the original parallel port. The protocol remains backward compatible with all the existing parallel port peripherals and printers.

IEEE 1284 defines five modes of data transfer, in either the forward direction (referenced to the PC), reverse direction, or both directions (bidirectional data transfer in a half-duplex format). The five modes have been distributed according to direction (see Table 5.1). The truly bidirectional modes are the EPP and the ECP modes, in which a single I/O cycle can transfer data to the external device while checking for a BUSY signal and generating the nSTROBE signal. EPP and ECP modes also allow the PC to read a byte of data from the external device, complete with all handshaking, in one I/O cycle.

All parallel ports can implement a bidirectional link using the nibble or byte mode. However, both these methods are software intensive. The data transfer software in a PC checks if the peripheral is ready (by probing the BUSY line), then places outgoing data onto the data lines and generates a strobe signal on one of the control lines.

Table 5.1   IEEE 1284 data transfer modes.
Forward direction only
•   Compatibility mode: Centronics or Standard mode
Reverse direction only
•   Nibble mode: four bits at a time using status lines for data
•   Byte mode: eight bits at a time using data lines, referred to as "bidirectional" port
Bidirectional
•   EPP mode: enhanced parallel port
•   ECP mode: extended capability port

Similarly, when reading data, the PC reads the data on the data lines (in byte mode) or a nibble on the status lines (in nibble mode) and then generates an acknowledge signal. All this is software intensive and limits the data transfer rates to between 50 and 150Kbps.

Newer PCs are equipped with I/O controllers with EPP and ECP capabilities and have hardware to assist with the data transfer. In EPP mode, a data byte can be transferred by a single IN or OUT instruction to the I/O controller, which then handles the handshaking and strobe signal generation. Clearly, data transfer rates on such a machine are then limited by the rate at which the instruction can be executed. Typically transfer rates around 1–1.75Mbps can be achieved easily on contemporary machines.

The 1284 standard not only specifies the data transfer protocol but also defines a physical and electrical interface. In addition, IEEE 1284 provides a method for the host and the peripherals to recognize the supported modes and negotiate the requested mode. What the IEEE 1284 standard does not do is specify how certain conditions are to be interpreted by the host. For instance, the standard does not specify how an EPP device signals that it has data to send or that an error condition has occurred. As you will see, the EPP protocol has the ability to signal an *address* or command. The standard does not provide a definition or meaning to any particular address or command, just a method to send. The 1284 standard is a low-level physical layer protocol.

The Enhanced Parallel Port

The Enhanced Parallel Port protocol was originally developed by Intel, Xircom, and Zenith Data Systems to provide a high-performance parallel port link that was backward compatible with existing parallel port peripherals and interfaces. Intel implemented this protocol in the 360SL I/O controller chip set. However, this was prior to the establishment of the IEEE 1284 committee.

Because EPP-capable parallel ports were available before the release of the IEEE 1284 standard there is small deviation in the pre-IEEE EPP ports from the 1284 EPP ports.

The EPP protocol defines four types of data transfer:

1. Data write cycle
2. Data read cycle
3. Address write cycle
4. Address read cycle

Data cycles are used to transfer data to and from peripheral devices and address cycles are used to exchange device addresses, control information, etc. An address cycle can be viewed as two different data cycles.

The EPP Protocol defines the SPP signal names, as shown in Table 5.2. Of the 17 SPP signals, EPP utilizes 14 signals for data transfer, handshake, and strobe purposes. The rest of the unused signals can be used by the peripheral designer for product-specific purposes.

The data write (Figure 5.1) cycles as follows:

1. Program executes an I/O write cycle to the EPP DATA port

2. The nWRITE line is asserted and the data is output on the parallel port data lines

3. The data strobe is asserted because nWAIT is asserted low

Table 5.2   EPP signal definitions.

SPP signal	EPP signal	Direction	EPP signal description
nSTROBE	nWRITE	Out	Active low. Indicates a write operation; high for a read cycle
nAUTOFEED	nDATASTB	Out	Active low. Indicates DATA_Read or DATA_Write operation in progress
nSELECTIN	nADDRSTB	Out	Active low. Indicates ADDRESS_Read or ADDRESS_Write operation in progress
nINIT	nRESET	Out	Active low. Peripheral reset
nACK	nINTR	In	Peripheral interrupt. Used to generate an interrupt to the host
BUSY	nWAIT	In	Handshake signal. When low, it indicates that it is OK to start a cycle (assert strobe); when high, it indicates it is OK to end the cycle (de-assert strobe)
D[8:1]	AD[8:1]	Bidirectional	Bidirectional address/data lines
PE	user defined	In	Can be used by peripheral
SELECT	user defined	In	Can be used by peripheral
nERROR	user defined	In	Can be used by peripheral

4. The port waits for the acknowledge signal from the peripheral (nWAIT de-asserted)
5. The data strobe is de-asserted and the EPP cycle ends
6. The ISA I/O cycle ends
7. nWAIT is asserted low to indicate that the next cycle may begin

The data read (see Figure 5.3) cycles as follows:

1. Program executes an I/O read cycle to the EPP DATA port
2. The data strobe is asserted because nWAIT is asserted low
3. The port reads the data bits and the data strobe is de-asserted
4. The port waits for the acknowledge from the peripheral (nWAIT de-asserted)
5. The EPP cycle ends
6. The ISA cycle ends

The address read and write cycles operate exactly the same way except that the nDATASTB signal is replaced by the nADDRSTB signal.

Figures 5.2 and 5.4 show the address write and read cycles. In essence, the EPP signals can be viewed as the signals of a general purpose microprocessor, with an 8-bit multiplexed address and a data bus. The control bus of this system could be seen as

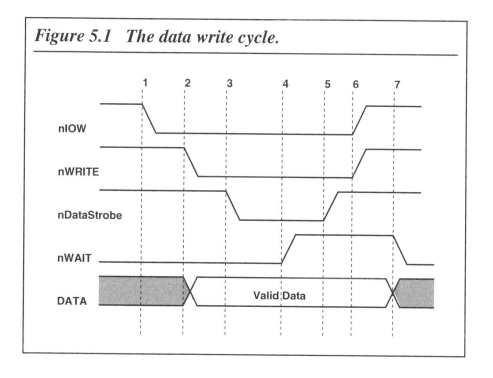

Figure 5.1   The data write cycle.

the collection of the six signals, composed of the nWRITE, nIOW, nIOR, nAddr-Strobe, nDataStrobe, and nWAIT signals. This is shown in Figure 5.5

EPP Registers

The register model for the EPP mode of operation is an extension of the IBM parallel port registers. As shown in Table 5.3, the SPP (Standard Parallel Port) register definitions include three registers offset from the port's base address: DATA port, STATUS port, and CONTROL port. The base address is the address where the data register is located in the PC's I/O space. This is commonly 0x378h or 0x278h for LPT1 and LPT2. The most common EPP implementations expand this to use ports not defined by the standard parallel port specification. This is shown in Table 5.3.

By generating a single I/O write instruction to base_address + 4 (0x27Ch for example), the EPP controller will generate the necessary handshake signals and strobes to transfer the data using an EPP Data_Write cycle. I/O instructions to the base addresses, ports 0 through 2, will cause the enhanced parallel port to behave exactly like the standard parallel port. This guarantees compatibility with standard parallel port peripherals and printers. Address cycles are generated when read or write I/O operations are generated to base_address + 3.

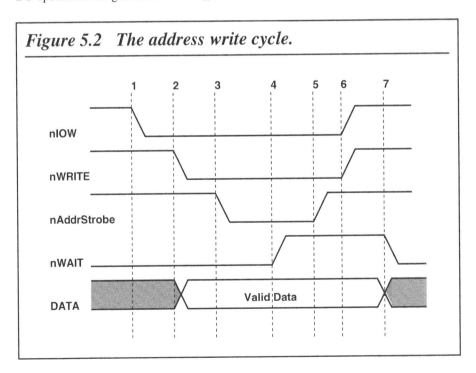

Figure 5.2   The address write cycle.

Figure 5.3   The data read cycle.

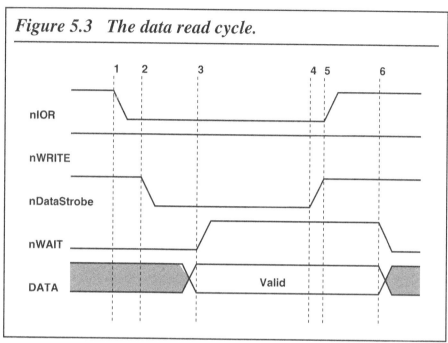

Figure 5.4   The address read cycle.

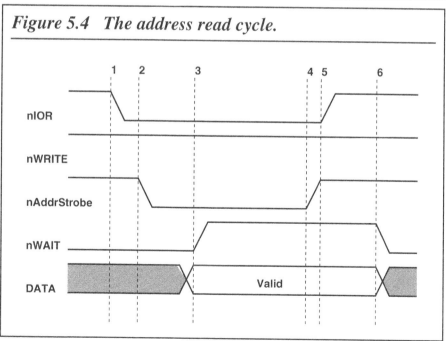

Ports 5 through 7 are sometimes used differently by various hardware implementations. Sometimes they are used to implement 16- or 32-bit software interfaces, or sometimes they are used as configuration registers. Sometimes they aren't used at all. Most controllers use these addresses to support 32-bit I/O instructions.

The ISA controller will intercept the 32-bit I/O and actually generate four fast 8-bit I/O cycles to registers +4 through +7. The first cycle will be to the addressed I/O port using byte 0 (bits 0–7), the second cycle will be to port+1 using byte 1, then port+2 using byte 2, and finally port+3 using byte 3. These additional cycles are generated by hardware and are transparent to the software. The total time for these four

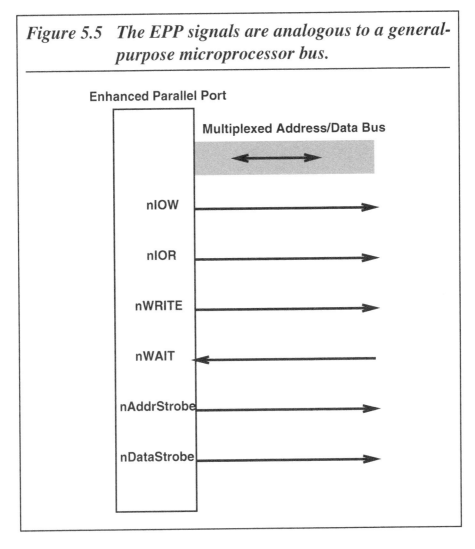

Figure 5.5 **The EPP signals are analogous to a general-purpose microprocessor bus.**

cycles will be less than four independent 8-bit cycles. This enables the software to use 32-bit I/O operations for EPP data transfer. Address cycles are still limited to 8-bit I/O. The ability to transfer data to or from the PC by the use of a single instruction is what enables EPP-mode parallel ports to transfer data at ISA bus speeds. Rather than having the software implement an I/O-intensive software loop, a block of data can be transferred with a single REP IO instruction. Depending upon the host adapter port implementation and the capability of the peripheral, an EPP port can transfer data from 500Kbps to nearly 2Mbps. This data transfer rate is more than enough to enable network adapters, CD-ROM, tape backup, and other peripherals to operate at nearly ISA bus performance levels.

The EPP protocol and current implementations provide a high degree of coupling between the peripheral driver and the peripheral. What this means is the software driver is always able to determine and control the state of communication to the peripheral at any given time. Block transfers and intermixing of read and write operations are thus easily accomplished. This type of coupling is ideal for many register-oriented or real-time controlled peripherals such as network adapters, data acquisition, portable hard drives, and other devices.

EPP BIOS Calls

Writing directly to the I/O controller (implementing the EPP) will always offer the fastest data transfer rates, but in some cases, you may wish to use the EPP BIOS calls for data transfer. Using BIOS calls allows data transfer without messing with the I/O

Table 5.3 EPP register definitions.

Name	Offset	Mode	Type	Description
SPP DATA port	+0	SPP	W	Standard SPP DATA port. No autostrobing
SPP STATUS port	+1	SPP	R	Reads the input status lines on the interface
SPP CONTROL port	+2	SPP	W	Sets the state of the output control lines
EPP ADDRESS port	+3	EPP	R/W	Generates an interlocked address read or write cycle
EPP DATA port	+4–7	EPP	R/W	Generates an interlocked data read or write cycle

controller chip. Another advantage of using the BIOS is that even though the host may not have an EPP-capable port, the software will still run using the EPP emulation mode, though at a highly reduced rate.

EPP BIOS calls provide a way to perform single I/O cycles as well as block transfers. Although the BIOS specifications cannot be reproduced here, I will describe some useful calls.

Installation Check This is used to test for the presence of an EPP port.

Set Mode This is used to set the operating mode of the EPP port.

Get Mode This is used to query the current operating mode of the EPP port.

Interrupt Control This is used to enable or disable the interrupt associated with the EPP port.

EPP Reset This is used to reset the peripheral connected to the EPP port.

Address Read This is used to perform an address read I/O cycle.

Address Write This is used to perform an address write I/O cycle.

Write Byte This is used to output a single byte via the EPP data port.

Write Block This is used to output a block of user-specified data from a defined buffer via the EPP data port data from a defined buffer via the EPP data port.

Read Byte This is used to read a single byte via the EPP data port.

Read Block This is used to input a stream of bytes into a user-defined buffer via the EPP data port.

Device Interrupt This is used to allow an EPP device driver to install an interrupt handler. Use this call whenever an EPP device interrupt occurs.

At this point in time, the EPP BIOS has been implemented by only a few companies. Most drivers will have to be written to use the registers directly. In the next section, I'll provide an example of high-speed digital I/O routines using EPP BIOS calls.

The full proposed specification for the EPP BIOS can be obtained from the IEEE P1284.3 draft specification.

High-Speed Digital I/O using EPP

Figure 5.6 shows the schematics for a simple DATA and ADDRESS I/O expansion port. The circuit shows two latches and two buffers and a simple decoding circuit. The circuit can be easily understood using the timing diagrams for the DATA and ADDRESS transfers described earlier in this chapter.

The latches are used to store a DATA byte as well as an ADDRESS byte. The buffers are also read as DATA input as well as ADDRESS input. In this simple circuit, the BUSY signal is generated from the combination of the nDATASTB and the nADDR-STB signals. In a more complex circuit (e.g., a circuit using a processor), the BUSY signal could be generated when the processor actually reads the data (or the address) so as to ensure that data or address is not lost.

Listing 5.1 shows the routines to implement high-speed digital I/O using EPP BIOS calls. The code is commented so that you can modified it for a given task.

Programming the EPP Controller Chip

Listing 5.1 uses the EPP in the previous section with the help of BIOS routine calls. The use of BIOS routines to access and communicate through the EPP of the PC is useful because it allows data transfer without the need for any knowledge of the underlying hardware details. All PCs are equipped with EPP controller chips (generally, the EPP is a part of a big I/O controller IC that contains a floppy controller and serial port controller as well as the EPP controller). Various manufacturers have implemented various functions in their respective chips and it is very difficult to develop a single piece of code that can be used to program all the EPP controllers directly.

Recall from Table 5.3 that the EPP port map consists of at least five registers: the SPP DATA port, SPP STATUS port, SPP CONTROL port, EPP ADDRESS port and the EPP DATA port. To be able to use the EPP ADDRESS and the EPP DATA ports, the EPP controller chip must be programmed into the EPP mode of operation. The default mode of operation of the parallel adapters in the PC is the standard parallel adapter mode (the SPP mode). To change the mode of operation, the controller chips have some sort of Extended Configuration Register (ECR). There seems to be a general understanding among the manufacturers that the ECR register would be placed at an address offset of 0x402h from the base address of the parallel adapter. If the base address of the parallel adapter is 0x378h, then the ECR will be accessible at address 0x77Ah.

By writing to the ECR register, the user can select one of the many modes of operation that the particular EPP controller supports. This has created some confusion, because the various manufacturers have implemented arbitrary mode selection bytes. As an example, Standard Microsystems Corporation, a major manufacturer of multi-mode parallel port and floppy disk controllers, offers the FDC37C665/66 chip. This

Figure 5.6 Digital I/O prototype circuit.

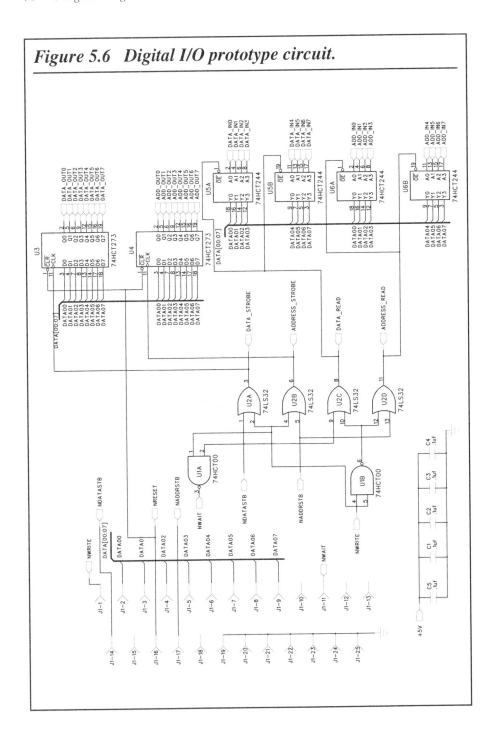

chip has an ECR at an address offset of 0x402 and it requires a mode selection to be byte written into the ECR. The various mode selection bytes are shown in Table 5.4.

National Semiconductor is another major supplier of multimode controller chips. Table 5.5 is the description of ECR mode selection for National Semiconductor's PC87332 chip.

Tables 5.4 and 5.5 show that the two chips are not identical, and thus it difficult to write common code to program the EPP controller chip.

However, in spite of these differences, there seems to be a similarity in accessing the various ports for all these chips. If you want to program the EPP controller chip directly, you must have the details of the particular chip used in your system.

Listing 5.1 Routines for high-speed I/O using EPP BIOS calls.

```
/*epp_io.c*/
/*Set of Routines for Digital I/O using the Enhanced Parallel
Port. The routines invoke EPP BIOS calls.
Uses Turbo C 2.0 or better
Typical results:
Block transfer rates: 65 Mbytes in 73 secs on 486/66
*/

#include<dos.h>
#include<time.h>
#include<stdio.h>

#define FALSE 0
#define TRUE 1

void far (*pointr)();

int epp_config(void);
int epp_write_byte(unsigned char tx_value);
int epp_write_block(unsigned char *source_ptr, int count);
int epp_read_byte(unsigned char *rx_value);
int epp_read_block(unsigned char *dest_ptr, int count);

int epp_config(void)
{
    unsigned char temp_ah, temp_al, temp_cl, temp_ch;
    _AX=0x0200;
    _DX=0;
    _CH='E';
```

Listing 5.1 (continued)

```
   _BL='P';_BH='P';
   geninterrupt(0x17);
   temp_ah=_AH;
   temp_al=_AL;
   temp_ch=_CH;
   temp_cl=_CL;

   if(temp_ah != 0) return FALSE;
   if(temp_al != 0x45) return FALSE;
   if(temp_ch != 0x50) return FALSE;
   if(temp_cl != 0x50) return FALSE;

   pointr = MK_FP(_DX , _BX);
   _AH=1;
   _DL=0;
   _AL=0x04;
   pointr();
   temp_ah=_AH;
   if(temp_ah != 0) return FALSE;
   return TRUE;
}

int epp_write_byte(unsigned char tx_value)
{
   unsigned char temp_ah;

   _AH=7;
   _DL=0;
   _AL=tx_value;
   pointr();
   temp_ah=_AH;
   if(temp_ah != 0) {return FALSE;}
   return TRUE;
}


int epp_write_block(unsigned char *source_ptr, int count)
{
   unsigned char temp_ah;

   _SI=FP_OFF(source_ptr);
   _ES=FP_SEG(source_ptr);
   _AH=8;
   _DL=0;
   _CX=count;
   pointr();
```

Listing 5.1 (continued)

```c
   temp_ah=_AH;
   if(temp_ah != 0)
   {
       printf("\nBlock write timeout error"); return FALSE;
   }
   return TRUE;
}
int epp_read_byte(unsigned char *rx_value)
{
   unsigned char temp_ah;
   _AH=9;
   _DL=0;
   pointr();
   *rx_value=_AL;
   temp_ah=_AH;
   if(temp_ah != 0) {return FALSE;}
   return TRUE;
}
int epp_read_block(unsigned char *dest_ptr, int count)
{
   unsigned char temp_ah;
   _DI=FP_OFF(dest_ptr);
   _ES=FP_SEG(dest_ptr);
   _AH=0x0a;
   _DL=0;
   _CX=count;
   pointr();
   temp_ah=_AH;
   if(temp_ah != 0) {return FALSE;}
   return TRUE;
}

main()
{
   int  ret_value, ret_val;
   time_t start, end;
   unsigned char buf_out[10000], buf_in[10000], rx_in;

   clrscr();
   printf("Fast Digital I/O using the Enhanced Parallel Port");
   printf("\nUses EPP BIOS Calls\nDhananjay V. Gadre\nJuly 1996");
```

In spite of these differences, the common approach in programming the EPP chip directly, would involve the following steps:

1. Detect, if the system is equipped with any EPP controller chip.

2. Select the correct mode byte for the particular EPP chip, so that the EPP mode is selected.

3. Write to the EPP ADDRESS port or the EPP DATA port to perform the required EPP transfers.

The Extended Capability Port

The Extended Capability Port (ECP) was proposed by Hewlett Packard and Microsoft. The idea was to provide a high-performance, multiple-capability communication channel between the PC and advanced printer/scanner-type instruments. In this category fall such instruments as a multifunction FAX/scanner/printer devices.

Listing 5.1 (continued)

```
ret_value = epp_config();
    if(ret_value == FALSE) { printf("\nNo EPP"); exit(1);}
printf("\nEPP Present");
printf("\n\nWriting Data Byte");
ret_val = epp_write_byte(0xa5);
    if (ret_val == TRUE) printf("\nWrite Byte Successful");
    else printf("\nTimeout error");
printf("\n\nWriting Data Block");
start=time(NULL);
    for(ret_value=0; ret_value<1000; ret_value++)
ret_val=epp_write_block(buf_out, 52428);
end=time(NULL);
printf("\nTime taken= %d seconds for 50 Mbytes", end-start);

printf("\n\nReading Data Byte..");
ret_val=epp_read_byte(&rx_in);
    if (ret_val == TRUE)
      printf("\nRead Data Byte Successful, %x", rx_in);
    else printf("\nRead Data byte failed");
printf("\n\nReading Data Block..");
ret_val=epp_read_block(buf_in, 1000);
    if (ret_val == TRUE) printf("\nRead Data Block Successful");
    else printf("\nRead Data block failed");
}
```

Similar to the EPP protocol, the ECP also configures the signals of the parallel adapter into a bus structure. However, unlike the EPP, which uses the bus to transfer data and addresses to and from the peripheral device, the ECP defines a more powerful protocol that allows multiple logical devices to coexist on the parallel adapter. The ECP also allows real-time data compression while transferring data across the peripheral device. The data compression is referred to as Run_Length_Encoding (RLE). This feature is particularly useful for scanners and printers that need to transfer large raster data. To use the RLE feature, the peripheral as well as the host must support it. The ECP defines two types of data transfers between the host (the PC) and the peripheral in either direction:

Table 5.4   ECR mode description for the FDC37C665/66 chip.

Mode (bits 7, 6, and 5 of the ECR)	*Description*
000	SPP mode
001	PS/2 parallel port mode
010	Parallel port data FIFO mode
011	ECP mode
100	EPP mode
101	Reserved
110	Test mode
111	Configuration mode

Table 5.5   ECR mode description for the PC87332 chip.

Mode (bits 7, 6, and 5 of the ECR)	*Description*
000	SPP mode
001	PS/2 parallel port mode
010	Parallel port data FIFO mode
011	ECP mode
100	Reserved
101	Reserved
110	FIFO test mode
111	Configuration mode

- Data transfer
- .Command transfer

Note that the ECP transfers commands rather than mere addresses across the channel. This suggests a complex hardware and software hierarchy to interpret, decode, and execute the commands. The commands can be composed of logical channel addresses or the run length count.

The commands as well as data are eight bits and are transferred on the eight DATA port bits. The HostAck signal identifies whether the data on the DATA port lines is data or a command. The type of the command is identified by looking at the most significant bit (MSB) of the command byte. If the MSB is 1, the rest of the bits represent the RLE count (thus the RLE count is between 0 and 127). If the MSB is 0, the rest of the bits represent the channel address (the channel address could be any number between 0 and 127).

Electrical Interface for Advanced Adapters

The IEEE 1284 committee, besides defining the many exotic data transfer modes for the parallel adapter, has also done a great job in defining the electrical interface for the PC as well as the peripheral devices. It is hoped that the general anarchy that reigns in the physical implementation of the parallel adapters will one day be arrested and a standard compatible interface will emerge in the future. In fact, with many varieties of the CONTROL port and the STATUS port, it is probably impossible to guarantee operation of high-speed protocols such as the EPP and the ECP.

The IEEE defines two levels of electrical interfaces: level I and level II. Level I interface is intended for products and devices that do not need to operate at high speeds but only need to use the reverse data transfer capabilities. This level is expected to be used for short distances. For devices that need to transfer data at full EPP and ECP specifications and for longer distances, the level II interface is recommended. The level II interface requires the following of the drivers:

- Open circuit high-level voltage shall be less than 5.5V; low-level voltage shall be greater than –0.5V
- Steady-state high-level voltage shall be at least 2.4V at a load sink of 14mA.
- Steady-state low-level voltage shall be at most 0.4V at a load source of 14mA.
- The driver output impedance measured at the connector shall be 50±5 ohms measured at an output voltage of $(V_{oh} - V_{ol})/2$.
- The driver slew rate shall be between 0.05 to 0.40V/ns.

The receiver features are:

- The receiver high-level input threshold voltage shall not exceed 2.0V and the low-level input voltage threshold shall be at least 0.8V

- The receiver shall withstand voltage transients between –2.0 and 7.0V without damage or improper operation.

- The receiver shall provide input hysteresis of at least 0.2V but not more than 1.2V

- The low-level sink current shall not exceed 20μA at 2.0V and the high-level source current not more than 20μA at 0.8V

- Total input capacitance shall be less than 50pF

Driver and receiver ICs meeting the IEEE1284 level II specifications are being produced by Texas Instruments and National Semiconductor.

Additional Information

There is a lot of activity in the standards arena associated with the 1284 parallel port standard. Two committees of note are the IEEE P1284.3 and IEEE P1284.4 committees. The P1284.3 committee is developing a standard protocol to implement port sharing (multiple devices on one parallel port) and also a full Data Link layer protocol for the 1284 port. The P1284.4 committee is developing a transport protocol that implements the concept of Multiple Logical Channels for a single communication channel. This protocol is targeted at printing and multifunction devices that connect to the parallel port. For more information, you can check the web site `http://www.fapo.com/ieee1284.htm`. From this web site you can find more information on the parallel port standard, as well as information on the new standards under development.

Chapter 6

Analog to Digital and Digital to Analog

Modern computers handle information represented as discrete binary voltages. However, to influence and interact with the external world, the computer must produce results that are analog, or continuously varying. This interfacing of the computer to the real world is achieved through devices that convert voltages to and from the analog domain. These special devices are called Analog to Digital Converters (ADCs) and Digital to Analog Converters (DACs). ADCs and DACs map the infinite levels of analog voltages to a discrete, finite set of digital levels.

Figure 6.1 shows the quantization of a steadily increasing analog voltage (which varies from a minimum to a maximum voltage through infinite levels) to discrete voltage levels. The first column shows two discrete levels, the second column shows four discrete levels, and the last column shows eight levels. Each discrete level is represented by a number or code. In this figure, the voltage thresholds at which the digital code changes have been marked on the vertical axis in the upper row of plots. In all three cases, when the analog voltage exceeds the voltage threshold, the number representing the voltage also changes. In the first case, because there is only one voltage threshold, the digital code changes after the analog voltage exceeds this threshold. To represent two digital codes, you could use any bit. So, the corresponding digital code for the two voltage levels could be 0 and 1. In the second case, there are three thresholds, and four discrete levels can be associated with these levels. To represent these four levels, you need two bits, with 00, 01, 10, and 11 as the corresponding digital codes. In the last case, you have seven thresholds, and hence, eight levels, for which you need eight codes: 000, 001, 010, 011, 100, 101, 110, and 111.

As you increase the number of quantization levels, the digital representation becomes closer to the actual analog voltage. In Figure 6.1, the three types of conversions are called 1-bit, 2-bit, and 3-bit conversions, respectively. (The term *n*-bit indicates how many digital bits are used for the representation.) With *n* bits, you will have 2^n possible digital codes. Typically, engineers use 8-, 10-, 12-, 16-bit, etc. converters, depending upon the required resolution.

What are DACs?

DACs are electronic devices that convert a digital code to an analog output in the form of a current or voltage. Functionally, the DAC has *n* digital input lines and one output line that provides analog voltage or current. The analog output is proportional to the weighted sum of the digital inputs:

$$V_{out} = K \bullet \sum_{j=1}^{n} (b_j \bullet 2^{j-1})$$

where, K is a constant and b_j is the *j*th digital input, which could be 0 or 1.

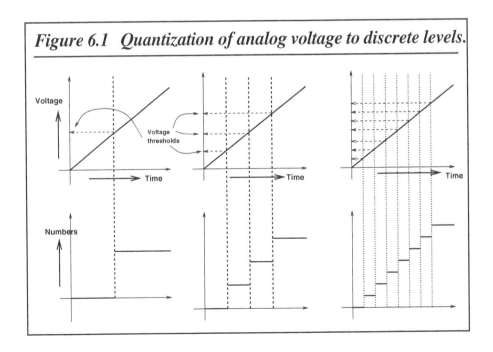

Figure 6.1 Quantization of analog voltage to discrete levels.

The various types of DAC implementations are:

- current switched,
- scaled resistor,
- R-2R ladder, and
- pulse width modulation.

The following sections discuss the various DAC implementations.

Current-Switched DAC

In a current-switched DAC, each of the digital (binary) input lines switches a current source proportional to the weight of the input bit. The digital bit in one logic state allows the current to flow, and in the other logic state stops any current flow. All input current sources are connected at a common point, and the sum of this current is the output for current output DACs, or this current is fed to a current-to-voltage converter (a resistor is the simplest current-to-voltage converter) to produce a voltage output.

Figure 6.2 illustrates a 4-bit, current-switched, voltage output DAC. The digital input bits are marked as B3 (MSB), B2, B1, and B0 (LSB). The most significant bit (B3) is controlling a current source of magnitude $8I$, and the least significant bit (B0)

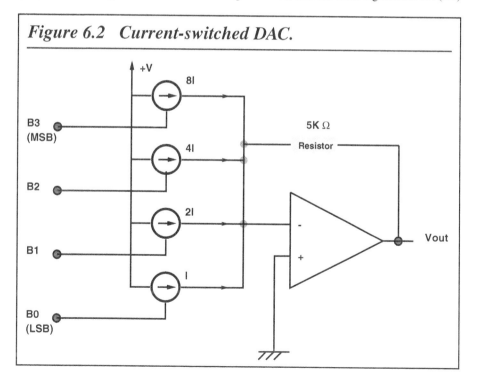

Figure 6.2   Current-switched DAC.

is controlling a current source of magnitude *I*. If any bit is 1, the corresponding current flows. If the bit is 0, no current flows. The current sources are connected at the inverting pin of an operational amplifier, which acts as a current summing junction. The sum of all the currents flowing into this junction pass into the 5Kohm resistor in the feedback path of the operational amplifier. The current flow through the resistor produces a voltage drop across the resistor, which appears at the output of the operational amplifier.

The voltage output of the operational amplifier when all the bits are 0 is 0V, and the voltage output for all the bits at logic 1 is $15 \times I \times 5,000$V. If *I* is set to 0.1mA, this voltage corresponds to 7.5V. Because the least significant bit (LSB) is controlling a current source of *I*, which is 0.1mA, the step size of the voltage output is 0.5V (i.e., when the input number changes by one, the output voltage changes by 0.5V).

If the input number is 0110, the sum of the currents is 6*I*, which is 0.6mA. This current flows into a 5,000-ohm resistor so the output voltage is 3V. If the number changes to 0111 (a change of one), the sum of the currents is 7*I*, which is equal to 0.7mA, and the corresponding voltage output is 3.5V. Thus the LSB of the DAC is equal to 0.5V. The process is similar for DACs of higher resolution (i.e., having more bits).

Scaled-Resistance DAC

Scaled-resistance DACs use an array of weighted resistances connected to the inverting input of an operational amplifier. The noninverting input is connected to ground. The junction of all the resistors offers a zero impedance to the currents flowing through the resistors. The sum of these currents flow into the feedback resistor connected between the output and the inverting terminal, producing a voltage drop across the resistor that appears at the output, similar to the switched-current DAC. The scheme is shown in Figure 6.3.

Figure 6.3 shows four bits B3 (MSB) to B0 (LSB) controlling a double-pole switch. The common terminal of the switch is connected to a resistor. The resistor for the B3 bit is the lowest value (10Kohm), which provides the largest current. One pole of the switch connects to the +ve supply voltage and the other pole connects to ground. If you assume that a logic 1 connects the common terminal to +ve supply and a logic 0 connects to ground, for a supply voltage of 10V, the currents produced for the four resistances would be 1.0, 0.5, 0.25, and 0.125mA. The maximum currents would flow when all the bits are at logic 1. The total current would be 1.0 + 0.5 + 0.25 + 0.125 = 1.875mA.

The voltage generated at this current at the output is the voltage drop across the 5Kohm feedback resistance and is 9.375V. Every LSB change in the input code would make the output voltage change by 0.625V.

R-2R Ladder DAC

The problem with the practical implementation of the scaled-resistance DAC is the difficult requirement of producing accurate resistances over a range of 1:2,000 for a 12-bit DAC. For lower resolution DACs, this may not be a problem, but with increasing resolutions, it is difficult to create these resistances. A better method is the R-2R ladder, which uses only two values of resistances: *R* and *2R* (50 and 100Kohm, for example).

Figure 6.4 shows a 5-bit R-2R ladder DAC. A reference voltage source supplies voltage to the ladder of resistances. The resistances with values *R* are connected in series and the last one is connected to ground. The resistances of value *2R* are connected at the junction of two *R* resistances, and these resistances are connected to a common point of a two-pole switch. One of the poles of the switch is grounded and the other pole is connected to the virtual ground of an inverting amplifier configuration (so called because the other terminal, the noninverting terminal, is grounded; because of the high input impedance and high gain amplifier, the voltage drop between the two input terminals of the op-amp is very close to zero).

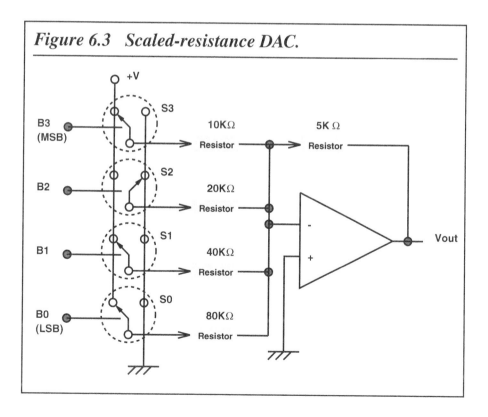

Figure 6.3   Scaled-resistance DAC.

Figure 6.5 shows the equivalent circuit of the R-2R ladder. The voltage at each R-R-2R resistor junction is half the voltage to the left and double the voltage to the right of the junction. Assume the reference voltage is 10V: $R = 50\text{Kohm}$ and $2R = 100\text{Kohm}$. With voltages at each of the R-R-2R junctions dropping by half, the currents flowing in each of the 2R resistance arms are 10/100, 5/100, 2.5/100, 1.25/100, and 0.625/100K amperes, which translates to 100, 50, 25, 12.5, and 6.25μA. Thus the currents decrease in ratios of increasing powers of two.

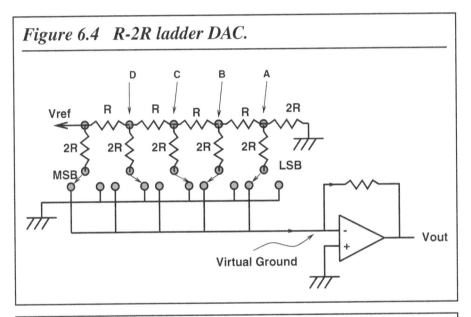

Figure 6.4   R-2R ladder DAC.

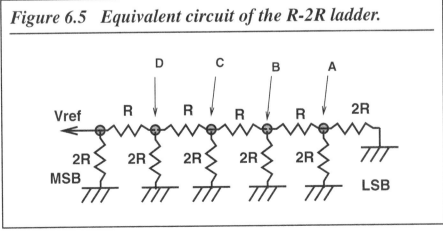

Figure 6.5   Equivalent circuit of the R-2R ladder.

Pulse Width Modulation (PWM) DACs

This type of DAC is usually available embedded within certain microcontrollers. The idea behind this type of DAC is that a digital wave of fixed frequency and variable duty cyle has a DC value proportional to its duty cycle. If this wave is properly low-pass filtered, the output of the filter will be the average voltage of the pulse wave. Typically, PWM DACs are used where the response time requirement is slow (e.g., the temperature control system for a water bath).

Multiplying DACs

As the name suggests, multiplying DACs are devices that allow multiplication of an analog voltage (or current) with a digital code. In all the previous examples, the DAC output was a function of either a reference voltage or a reference current. If, instead of generating this current or voltage internally, the DAC allows the user to supply it from the outside, the DAC output will be *I* × *inputbinarycode* or V_{ref} × *inputbinarycode*. The result is a multiplying DAC. All DACs do not allow the reference current or voltage to be supplied from an external source. DACs that allow this are specifically called multiplying DACs.

Popular DACs

Some common DACs are:

*   DAC0800
*   AD558
*   AD7548
*   MAX521

The following sections describe each popular DACs.

DAC0800

The DAC0800 series of DACs are monolithic, 8-bit, current output DACs. These devices offer 100ns settling time and can also be used as multiplying DACs with a 40:1 variation in the reference current. The full scale error is ±1 LSB. The operating supply voltage is ±4.5 to ±18V. The digital inputs can accept TTL signals directly (irrespective of the supply voltage). Figure 6.6 shows a typical operating scheme. The components in the figure produce 20Vp-p output.

To use this DAC, you need to connect the output of a suitable latch to the inputs of the DAC. Other components are suitably selected according to the DAC output voltage requirement.

The DAC has two current sink output pins Iout and Iout* (i.e., the current flow into the DAC). The output current is controlled by the input code as well as the reference current setup by the R_{ref} resistance and the external reference voltage source. In the block diagram these values are 5Kohm and 10V, respectively. This produces a reference current I_{ref} of 2.0mA. The total output current is $I_{tot} = (inputcode/256)I_{ref}$. The two output currents are generated such that $I_{out} + I_{out*} = I_{tot}$.

If I_{out*} is set to zero by connecting the Iout signal pin to ground, $I_{out} = I_{tot} = (inputcode/256)I_{ref}$. You can convert this current to voltage by connecting a suitable resistor between a +ve supply voltage and the current output pin.

AD558

AD558 is a complete voltage output DAC in two calibrated ranges. This device has an internal precision reference voltage source. An AD558 can operate with a widely varying supply voltage — between +5 and +15V — and has a direct microprocessor interface (so you don't need a latch to feed the DAC).

The block diagram in Figure 6.7 shows an input latch, which must be connected to the microprocessor data bus. The data is set up on the latch input lines and a latching pulse is used to latch the data. The latch outputs drive the 8-bit voltage output DAC. In case the data is to be driven from the outputs of an existing latch (e.g., the outputs of the DATA port of the parallel adapter), you can make the input latch of the AD558 transparent by connecting the input control pins to ground. The output pins of the DAC allow the user to select a 0–2.56V and a 0–10V output range.

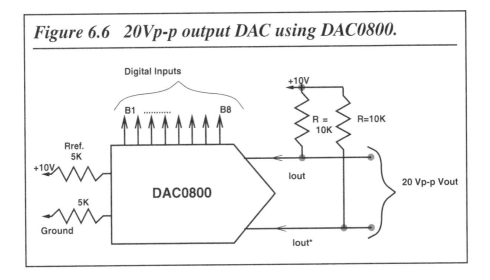

Figure 6.6   20Vp-p output DAC using DAC0800.

AD7548

AD7548 is a 12-bit voltage output DAC that can be directly connected to an 8-bit microprocessor data bus. The DAC is fabricated with CMOS technology, which operates at +5, +12, or +15V power supply voltages. Internally, the DAC is implemented using a R-2R ladder. The data to the DAC is loaded in two cycles. The data can be loaded such that it is left or right justified. It can also be loaded with the least significant byte first or the most significant byte first. (See the device data sheets for details of the various loading formats.) Figure 6.8 shows the format of an AD7548 connected to the parallel adapter. The voltage output of the DAC in this case is

$$V_{out} = V_{ref} \bullet (input code)/4096$$

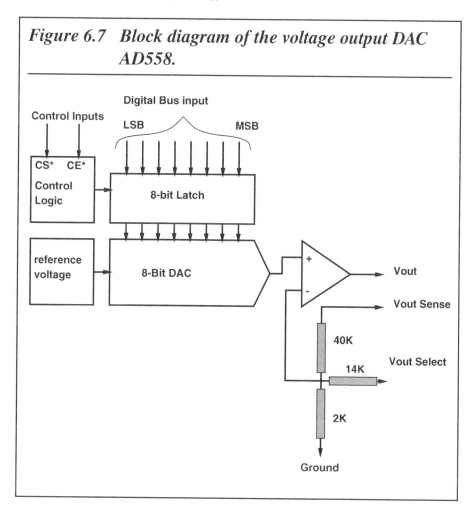

Figure 6.7   Block diagram of the voltage output DAC AD558.

where *inputcode* is a 12-bit number (between 0 and 4095) from the DATA port of the parallel port.

Figure 6.8 Circuit schematic of the AD7548 interface to the parallel adapter.

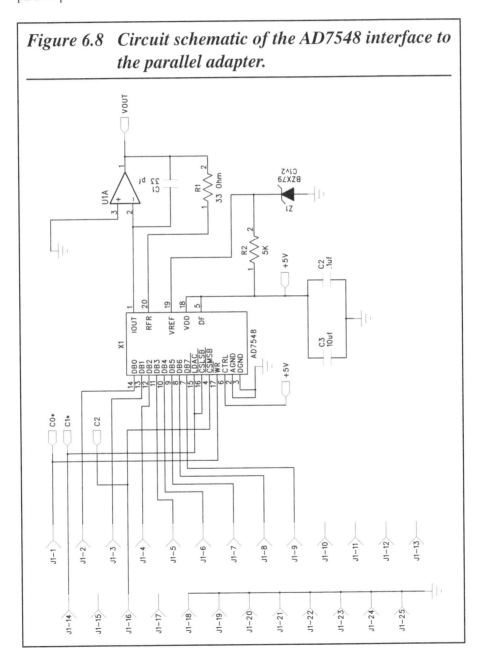

The reference voltage V_{ref} is provided by a 1.2V reference zener diode, so the maximum voltage output is approximately 1.2V. The LSB of the system is 1.2/ 4,096V, which is approximately 0.3mV. Listing 6.1 is a program to interface the 12-bit AD7548 ADC to the DATA port of the parallel adapter.

Listing 6.1   Interface for the 12-bit AD7548 ADC.

```c
/*pc_dac.c*/
/*Program to interface the 12-bit AD7548 ADC to the DATA port of the
  parallel adapter*/

#include <stdio.h>
#include <dos.h>
#include <conio.h>
#include <process.h>

#define BUF_LEN 1000

/*Global variables that store the addresses of two of three ports of the
  standard printer adapter*/
/*for interfacing the DAC to the parallel adapter we need the DATA
  port and the CONTROL port*/

unsigned int dport, cport;


main()
{
    /*the following array stores data to be transfered to the DAC
      connected to the DATA port*/
    unsigned int dac_array[BUF_LEN], count, in_temp;
    unsigned char temp;


    /*Get LPT1 port addresses */
    dport = peek(0x40,0x08);
    if(dport ==0)
        {
        printf("\n\n\nLPT! not available... aborting\n\n\n");
        exit(1);
        }
    printf("\nLPT1 address = %X", dport);
    cport = dport +2;     /* control port address */
```

MAX521

MAX521 is an octal, 8-bit, voltage output DAC with a simple two-wire serial interface to allow communication between multiple devices. The MAX521 operates from a single +5V supply voltage, and the voltage outputs can swing rail-to-rail. The DAC has five reference inputs that can be set to any voltage between the supply voltage levels. The MAX521 is available in a 20-pin DIP package. This DAC is described in complete detail, including an interface to the parallel adapter, in Chapter 8.

Listing 6.1 (continued)

```
/*this statement puts all the CONTROL port signals to logic 1*/
outportb(cport, 0x04);

/*setup a loop to transfer the required data points*/
for(count=0; count<BUF_LEN; count++)
   {
   /*The DAC data is stored in the integer variable such that the
     least significant 12 bits of the 16 bits contain the DAC data.
     The DATA port is 8-bits wide so the 12-bits data is transferred
     in two passes. In the first pass, the MSB data is transferred,
     which is 4 bits. In the next pass, the lower 8 bits are transferred
     and the LDAC signal is activated
   */

   in_temp = dac_array[count];
   in_temp = in_temp>>8; /*transfer the higher byte in the lower position*/
   temp=(unsigned char) in_temp;
   /*copy the low byte of the int into a char variable*/

   outportb(dport, temp);    /*output it to the DATA port*/
   temp = inportb(cport);
   temp = temp & 0xfb; /*make C2 low, to pulse CSMSB* low*/
   outportb(cport, temp);
   temp = temp | 0x01;   /*pulse C0* low to generate WR* */
   outportb(cport, temp);
   temp = temp & 0xfe;              /*make C0* high again*/
   outportb(cport, temp);

   temp = temp | 0x04;              /*make C2 high again*/
   outportb(cport, temp);

   in_temp = dac_array[count];
   in_temp = in_temp & 0x0f;
   temp=(unsigned char) in_temp; /* put the lower byte of the integer
                                    variable in a char variable*/
```

What are ADCs?

Analog-to-Digital Converter (ADC) devices convert analog signals (in the form of voltage or current) to numbers that the computer can handle. Functionally, the ADC has *n* output (digital) bits for an *n*-bit converter. The input bits indicate the digital equivalent of the analog quantity at the ADC input. Besides these output lines, the ADC has a Start Conversion (SC) signal that signals the ADC to begin conversion. Unlike DACs, most ADCs are slow compared to the digital circuits that operate them. The control circuit issues the SC signal to the ADC. After the end of the conversion, the ADC signals to the control circuit that the conversion is over and that the data can be read. This is indicated using the End Of Conversion (EOC) signal. To allow the output of the ADC to be connected on a common bus, there may be an output enable (OE) signal, which the control circuit must activate for the converted data to appear on the output bits.

Some typical ADC signals are shown in Figure 6.9. The various types of ADCs are described in the following sections. Each type offers various conversion speeds, noise performance levels, resolutions (some ADC types do not come with large resolution), and costs, and each type is suited for a different application.

Flash ADCs

Flash ADCs are the fastest ADCs, and they are also the easiest to understand. For an *n*-bit converter, a flash ADC requires (2n − 1) comparators. For a 10-bit flash converter, you

---

Listing 6.1 (continued)

```
    outportb(dport, temp); /*load the DATA port with the lower byte for
                    the DAC*/
    temp = inport(cport);temp = temp | 0x02;
    outportb(cport, temp); /*make C1* low so that CSLSB* and LDAC are low*/

    temp = temp | 0x01; /*pulse C0* low to generate WR* */
    outportb(cport, temp);

    temp = temp & 0xfe;            /*make C0* high again*/
    outportb(cport, temp);

    temp = temp & 0xfd;            /*make C1* high again*/
    outportb(cport, temp);
    }
}
```

must have 1,023 comparators on the chip, which is quite demanding not only in terms of fabricating sheer number of components but also in terms of power consumption.

Figure 6.10 shows the block diagram of a 2-bit flash ADC. For a 2-bit ADC, you need three comparators, as shown. The comparators have inverting and noninverting inputs. All the noninverting inputs are shorted together and the input voltage is connected here. The three inverting inputs are connected to a resistor network of resistor values $R/2$ and R. For a 3V reference driving the resistor network, the resistors are arranged such that the three comparators get a voltage equal to 2.5, 1.5, and 0.5V from the top comparator to the bottom comparator, respectively.

The three outputs of the comparators feed a digital combinational circuit, which produces two output bits from these three input bits. Table 6.1 shows the outputs of the comparators and the outputs of the ADC for hypothetical values of the input voltage.

Typical speeds for flash ADCs are 10 to 1,000 Msamples/s, and typical resolutions are 8-bit and 10-bit. Flash ADCs are used for applications such as frame grabbers and digital scopes.

Sampling ADCs

The sampling ADC, also called the successive approximation ADC, is a closed loop analog-to-digital converter. A sampling ADC uses the successive approximation algorithm to implement the conversion. The successive approximation algorithm begins by assuming the input voltage is exactly half the range of the ADC. The control unit

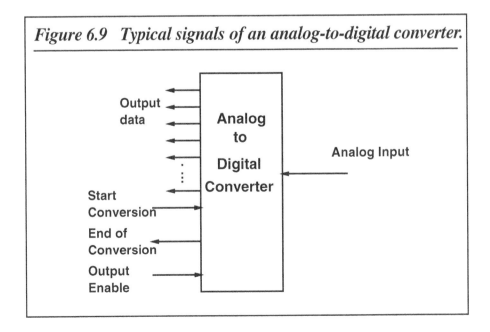

Figure 6.9   Typical signals of an analog-to-digital converter.

Table 6.1   Flash ADC input and output values.

Vin (V)	Comparator output	ADC output
2.8	111	11
2.0	011	10
1.0	001	01
0.25	000	00

Figure 6.10   A 3-bit flash-type analog-to-digital converter.

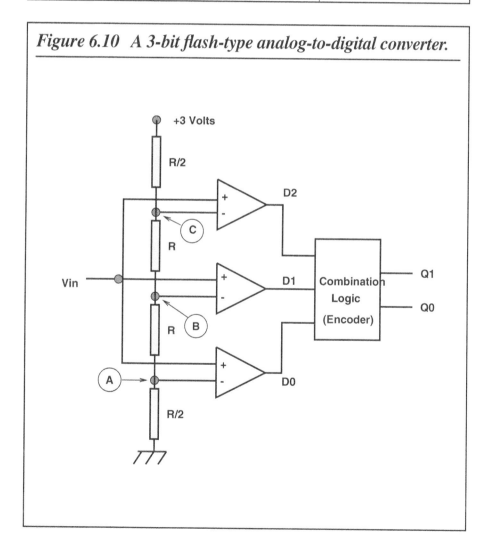

produces the code for this assumption and feeds this code to a DAC. The DAC output is compared with the input voltage using a comparator. If the comparison indicates the initial assumption was wrong (i.e., the assumed voltage was higher or lower than the input voltage), the control unit adjusts the previous assumption and initiates another comparison. This process continues recursively until the least significant bit is within tolerance. The end value is the code representing the input voltage, and this code is output on the output lines of the ADC.

Figure 6.11 shows the block diagram of a sampling ADC. Because the input voltage can change during the conversion process, an error could occur in the result. To minimize the errors, the ADC is equipped with a device called a sample-and-hold amplifier, which samples the input voltage and stores this voltage on a capacitor, disconnecting the input during the conversion process. This ensures that even though the input voltage may change, the voltage presented to the ADC comparator remains fixed.

Integrating ADCs

Integrating ADCs are slow but accurate converters useful for voltage measurements in noisy environments. Integrating ADCs are typically used in voltmeter and similar instrumentation applications.

The idea of an integrating ADC is to charge a capacitor with a current proportional to the input analog voltage to a fixed time *T*. After time *T* elapses, the capacitor is discharged at a constant rate through a current sink. The time it takes for the capacitor to

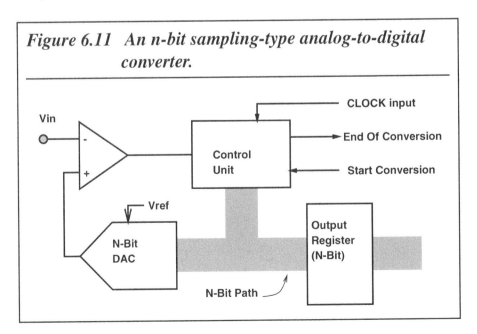

Figure 6.11 An n-bit sampling-type analog-to-digital converter.

discharge to 0V is proportional to the input voltage. A counter counts a stable clock pulse during the time the capacitor is discharging through the current sink. The count accumulated by the counter is the ADC output.

The advantage is that the component's absolute accuracy is not important — only the short-term stability is important. The capacitor charging time T should be a multiple of the power line time period.

Figure 6.12 shows the charging and discharging of the capacitor for two input voltages. The higher input voltage charges the capacitor to a higher voltage, and hence, the capacitor takes a longer time to discharge compared to the time required to discharge a lesser input voltage. The capacitor discharges into a current sink, so the rate of discharge (i.e., the slope of the discharge graph) will be same for the two cases.

Figure 6.13 shows the block diagram of the integrating ADC. The control circuit resets the counter at the beginning of the discharge cycle. The zero voltage switch detects when the capacitor voltage becomes zero. At that moment, the zero voltage switch signals to the control circuit, which stops the counters.

Figure 6.12   Charge and discharge of the capacitor in an integrating ADC.

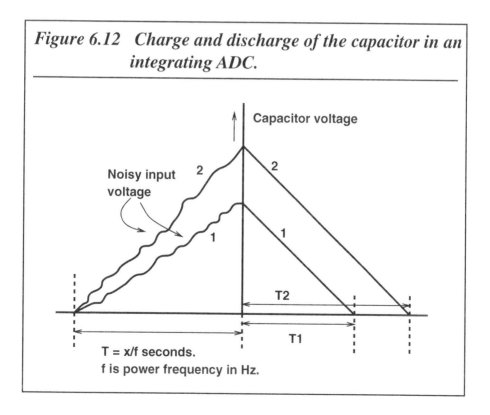

Popular ADCs

Some popular and widely available ADC devices are:

* ADC0804
* MAX158
* MAX186
* MAX111

The following sections discuss these popular ADCs. In these examples, you will perhaps see a bias towards Maxim ADCs. This bias is not necessarily because Maxim makes the best ADCs, but because Maxim is very generous in providing samples. For certain critical applications (such as applications demanding high resolutions of 16 bits, high speeds of 500Ksamples/s, low noise, and low cost) Analog Devices makes some excellent converters. Similarly, Harris, Linear Devices, National Semiconductors, Texas Instruments, and Burr-Brown are some of the major ADC manufacturers. It is useful to investigate all the options before settling on any particular device. The four devices I have chosen represent a reasonably wide selection of available ADCs.

ADC0804

ADC0804 is a very old (some call it obsolete) industry standard, single-chip ADC. Originally produced by National Semiconductors, the ADC0804 is also second sourced by Harris.

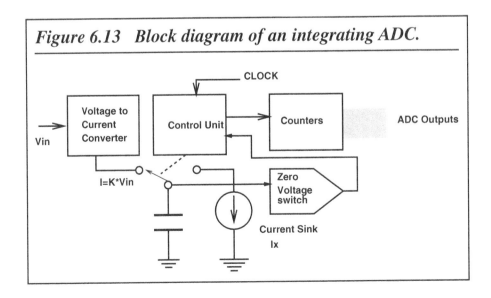

Figure 6.13   Block diagram of an integrating ADC.

The ADC0804 is an 8-bit successive approximation ADC. It is available in a 20-pin DIP package. The ADC0804 works at +5V supply voltage and consumes only 1.5mA of current, ideal for low-power portable data acquisition applications.

The successive approximation principle involves comparing the input voltage with an internally generated voltage until a best match is found. The internal voltage is generated across a tapped resistor array connected to the internal voltage reference source. The most significant bit is tested first, and eight subsequent comparisons provide an 8-bit binary code. After the last comparison, this 8-bit code is transferred to an output latch and can be read by manipulating the $\overline{RD}$ and $\overline{CS}$ signals. The ADC has an internal oscillator but does not include resonating components, which must be supplied externally. Otherwise, an external clock signal can be connected to the CLK IN input pin. The ADC operates with a maximum clock signal of 640KHz. To use the internal oscillator of the ADC, you must have an external resistor and a capacitor of the appropriate values.

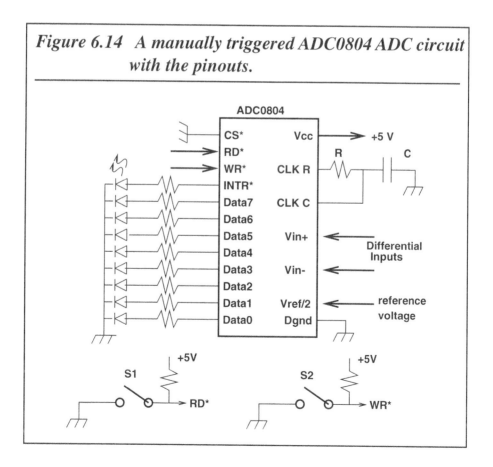

Figure 6.14   A manually triggered ADC0804 ADC circuit with the pinouts.

Figure 6.14 shows the block diagram of a circuit to convert analog voltage at the input of the ADC and to display the result on LEDs. The ADC conversion process is triggered by an on–off switch. The data is output to the LEDs when the other switch is operated.

To convert an input voltage into an 8-bit digital code, the user needs to manipulate the WR*, CS*, and RD* signal inputs. These signals are normally (i.e., when the ADC is not being used) set to a logic high state. The process of converting an analog voltage to a digital code involves the following steps:

1. trigger a conversion process;

2. wait for the conversion process to end;

3. read the resultant digital code.

The CS* signal (the Chip Select signal of the ADC) can be thought of as a master control signal that must be enabled in the conversion as well as the readout process. Together with CS*, the WR* signal is used in the conversion process. To initiate conversion, the CS* is put to a logic low state. This is followed by a logic high-to-low transition on the WR* signal pin, which puts the ADC in a RESET condition. The output signal INTR* is set high to indicate the initiation of the conversion process. Conversion starts after 1–8 clock cycles have elapsed after the WR* signal transits to logic high. Thereafter, the ADC takes the requisite number of clock cycles to complete the conversion. At the end of the process, the INTR* signal makes a high-to-low transition. The resultant code is then available and can be read from the eight DATA bits, or a new conversion process can be initiated.

To read the digital data, RD* is set low by the user (with CS* at logic zero). The digital data appears at the data output pins at this moment and can be suitably read. This also sets the INTR* signal to logic high, which is the default level of INTR*.

Figure 6.15 shows the timing diagram of a conversion and readout process using an ADC0804.

MAX158

MAX158 is an 8-channel, 8-bit fast (3µs conversion time) sampling ADC system. It includes an internal reference voltage source and a built-in sample-and-hold amplifier. All the features of the MAX158 and a PC parallel adapter interface are described in a later chapter.

MAX186

MAX186 is a 12-bit, 8-channel serial ADC system with built-in voltage reference, internal sample, and hold amplifier and multiplexer. The ADC offers various modes of operation, such as single-ended conversion, differential conversion, and sleep mode. The maximum current conversion is 2mA and 100µA during low-power modes (sleep mode). This ADC is also described in a later chapter, complete with a working schematic and code for a parallel adapter PC interface.

MAX111

MAX111/MAX110 is a serial 14-bit, dual-channel ADC from Maxim. The MAX111/MAX110 ADC uses an internal auto-calibration technique to achieve 14-bit resolu-

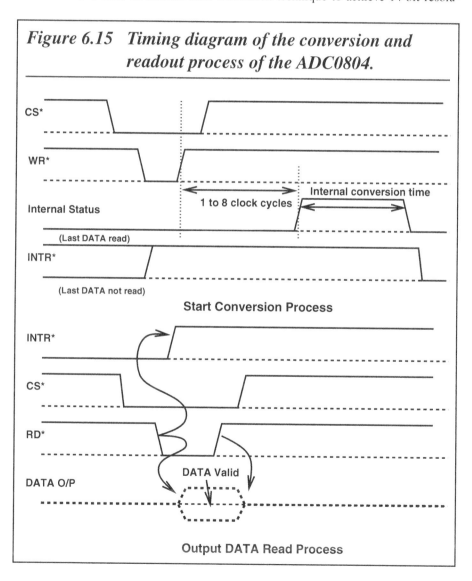

Figure 6.15   Timing diagram of the conversion and readout process of the ADC0804.

tion without any external component. The ADC offers two channels of ADC conversion and operates with 650µA current, thus making it ideal for portable, battery-operated data acquisition operations. We plan to use this interface with a battery-operated astronomical photometer to record night sky brightness and other observations together with a laptop computer.

MAX111 operates from a single +5V power supply and converts differential signal in the range of ±1.5V or differential signals in the range of 0–1.5V.

MAX111 can operate from an external or an internal oversampling clock, which is used for the ADC conversion. To start a conversion, digital data is shifted into the MAX111 serial register after pulling the $\overline{CS}$ low. $\overline{CS}$ can only be pulled low when $\overline{BUSY}$ is inactive. MAX111 has a fully static serial I/O shift register that can be read at any serial clock (SCLK) rates from DC to 2MHz. Input data to the ADC is clocked in at the rising edge of the SCLK and the output data from the ADC (conversion result) is clocked out at the SCLK falling edge and should be read on the SCLK rising edge.

The data clocked into the ADC determines the ADC operation. That data could initiate a new conversion, calibrate the ADC, perform an offset null, change an ADC channel, or change the oversampling clock divider ratio. The format of this control word is shown in Table 6.2.

Figure 6.16 shows the MAX111-to-PC interface. The ADC draws power from an external battery source stabilized by a local 78L05 regulator.

Listing 6.2 shows a program that will acquire and display samples from the MAX111. The interface works by monitoring the status of the BUSY* signal, which indicates if the ADC is busy with a conversion. A 0 on this pin indicates that the ADC is still converting. The program reads the status of BUSY* on the S6 pin of the printer port (STATUS port bit 6). When the program finds BUSY* at logic 1, it pulls the CS* signal of the ADC low to start a new conversion process. It then generates 16 clock pulses on the D0 pin of the printer port, which is connected to the SCLK signal pin of the ADC. Synchronized to these pulses, the program generates a serial bit stream on pin D7 of the printer port connected to the Din pin of the ADC. This bit stream contains the control word with the format described in Table 6.2. Output data from the ADC is clocked out on the Dout pin on the falling edges of the SCLK pulses. The program reads this data on the S7* signal pin of the printer port. The CS* signal is pulled up after the 16 clock pulses are generated. The ADC pulls its BUSY* signal low while the conversion is in progress. The conversion time depends upon the SCLK frequency and the format of the control word. In this circuit, the internal RC oscillator is used for the conversion clock. The converted data is clocked out in the next round of the clocking sequence by the ADC.

Figure 6.17 shows the timing diagram of a typical conversion and readout sequence recorded on a logic analyzer.

Table 6.2   ADC control word.

Bit #	15	14	13	12	11	10	9	8
	No-op	NU	NU	CONV4	CONV3	CONV2	CONV1	DV4
Bit #	7	6	5	4	3	2	1	0
	DV2	NU	NU	CHS	CAL	NUL	PDX	PD

Bit name	Function
No-op	If this bit is 1, the remaining 15 bits are transferred to the control register and a new conversion begins when CS* returns high
NU	Not used, should be set low
CONV1–4	Conversion time control bits
DV4–2	Oversampling clock ratio control bits
CHS	Input channel select; logic 1 selects channel 2, low selects channel 1
CAL	Gain calibration bit; a high bit selects gain calibration mode
NUL	Internal offset null bit; logic high selects this mode
PDX	Oscillator power-down bit, selected with logic high
PD	Analog power-down bit, selected with logic high

Figure 6.16 PC interface for the MAX111 ADC.

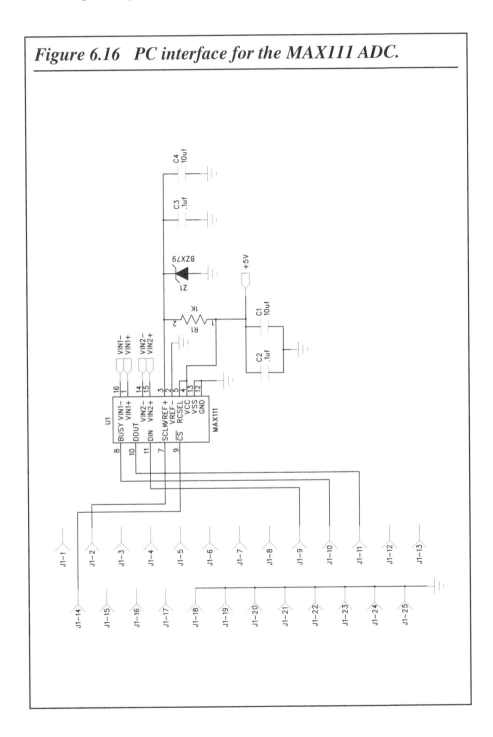

Figure 6.17   Timing diagram of the conversion and readout process of the MAX111.

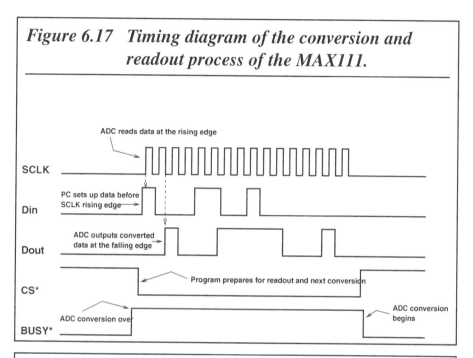

Listing 6.2   Interface for the 14-bit MAX111 ADC.

```
/*max111.c*/
#include<stdio.h>
#include<dos.h>
#include<time.h>
#include <conio.h>
#include <process.h>

/* Interface for 14 bit MAX111 */

/*
   Printer adapter pin usage
   --------------------------
   D7 (9)  = Din
   D0 (2)  = SCLK
   C1* (14)= CS*
   S7* (11)= Dout
   S6  (10)= BUSY
*/

#define TRUE 1
#define FALSE 0
```

Listing 6.2 (continued)

```
/* Global variable has address of DATA port of LPT1 */
 unsigned int dport_lpt1; /* data port address*/

int chk_lpt(void);                    /*Check if LPT1 is present*/
void disble_adc(void);    /*Pull ADC CS* high to disable ADC*/
void enable_adc(void);     /*Pull ADC CS* low to enable ADC*/
void chk_adc_status(void); /*Check ADC for End of Conversion*/
unsigned int read_adc(unsigned int prog_word);


unsigned int read_adc(unsigned int prog_word)
{
   unsigned char temp2, temp1, temp3;
   unsigned int temp_val, adc_val, out_val;

   out_val=prog_word;
   chk_adc_status();
   enable_adc();
   adc_val=0;

   for(temp2=0; temp2<16; temp2++)
   {
       temp1=inportb(dport_lpt1);
       temp1=temp1 & 0x7e;

       temp_val=out_val << temp2;
       temp_val=temp_val >> 8;
       temp_val=temp_val & 0x0080;

       temp1=temp1 | (unsigned char)temp_val;
       outportb(dport_lpt1, temp1);        /*send data*/
       temp1=temp1 | 0x01;
       outportb(dport_lpt1, temp1);        /*send Sclk*/

       temp3=inportb(dport_lpt1+1) & 0x80; /*read data*/
       temp3=temp3 ^ 0x80;
       temp_val=(unsigned int)temp3;
       temp_val=temp_val<<8;
       temp_val=temp_val & 0x8000;
       temp_val=temp_val>>temp2;
       adc_val=temp_val | adc_val;

       temp1=temp1 & 0xfe;
       outportb(dport_lpt1, temp1);
   }
```

Listing 6.2 *(continued)*

```
   outportb(dport_lpt1+2, (inportb(dport_lpt1+2) & 0xfd) );
   return adc_val;
}

void chk_adc_status(void)
{
   unsigned char tempa;

   tempa=inportb( dport_lpt1+1);
   tempa=tempa & 0x40;
    while(!tempa)
    {
           tempa=inportb(dport_lpt1+1);
           tempa=tempa & 0x40;
    }

}

void disable_adc(void)
{
   unsigned char tempx;

   tempx=inportb(dport_lpt1+2);
   tempx=tempx & 0xfd;
   outportb(dport_lpt1+2, tempx);
}

void enable_adc(void)
{
   unsigned char tempy;

   tempy=inportb(dport_lpt1+2);
   tempy=tempy | 0x2;
   outportb(dport_lpt1+2, tempy);
}

int chk_lpt(void)
{
   /*Get LPT1 port addresses */
   dport_lpt1 = peek(0x40,0x08);
   if(dport_lpt1 == 0)
   return FALSE;

   /* else return TRUE */
   return TRUE;
}
```

Listing 6.2 (continued)

```
main()
{
    float final_vol;
    unsigned int adc_val;
    unsigned char temp1;
    clrscr();

    /*Check if Printer port is present*/
    if( chk_lpt() == FALSE)
    {
        printf("\nPrinter port not available... aborting"); exit(1);
    }
    printf("\nPrinter port is at %X hex", dport_lpt1);

    /*Disable the ADC*/
    disable_adc();

    /*set clock and data low*/
    temp1=inportb(dport_lpt1);
    temp1=temp1 & 0x7e;
    outportb(dport_lpt1, temp1);

    printf("\nWaiting for the ADC to be ready...");
    chk_adc_status();
    printf("\nADC is ready for conversion...");

    /*perform offset correction for channel 1 */
    adc_val=read_adc(0x8c8c);

    /*perform gain claibration*/
    adc_val=read_adc(0x8c88);

    /*perform offset null*/
    adc_val=read_adc(0x8c84);

    /*Now convert indefinately on channel 1*/
    for(;;)
    {
        adc_val=read_adc(0x8c80);
        adc_val=adc_val & 0x3fff;
        final_vol=(float)adc_val;
        final_vol=(final_vol/16384)*1.2235; /*Vref=1.2235V*/
        printf("\nADC Value=%2.4f Volts", final_vol);
        delay(1000);
    }
}
```

## Chapter 7

Measuring Time and Frequency

For many applications, you must measure the frequency of some signal or the time interval between two activities. This chapter describes how to provide frequency and period measurement capability to a PC using the parallel port. The examples in this chapter use a low-resolution 32.768KHz crystal commonly used in electronic clocks. You can adapt the examples in this chapter to your own specific needs.

If you adapt these examples for your own applications, be aware that you may need to adjust the hardware and software components accordingly. You may find, for instance, that the frequency meter should have a higher resolution or a wider length than what is described.

Figure 7.1 is the block diagram of a simple frequency counter. The frequency counter block diagram shows the input frequency entering the box labeled Input Amplifier + Waveshaper. This circuit is a suitable amplifier of suitable gain (such that even signals with feeble amplitude can be recorded) with input protection (for signals with voltages above the operating voltages of the circuit). For nondigital input signals, the circuit has a waveshaper that produces digital signals from the input signals. Typically a Schmidt trigger circuit would be used at this point. The output of the input amplifier and waveshaper is a neat digital signal with a measurable frequency. This signal is fed into the second box labeled Controlled Gate. As the name implies, the controlled gate circuit regulates the propagation of the incoming signals to the counter chain under control of the gate control signal from the time base generator. In its simplest form, the controlled gate is nothing but an AND gate. One input of the AND gate is connected to the incoming signals and the other input is connected to the gate control signal. As long as the gate control signal is low (which you assume is the inactive state of the control signal), the input signal is blocked by the AND gate (i.e., the output of the AND gate is always low). When the gate control signal goes high, the AND gate allows the incoming signal to progress into the counter chain circuit.

The time base generator and the control circuit are at the heart of the frequency counter. The time base generator generates the control signal for the control gate. Consider that you want to measure the frequency to a resolution of 1Hz. A gate control signal of one second will open the gate for one second and the counters will measure the frequency, which would be resolved to 1Hz. For a resolution of 10 Hz, a gate control signal of 0.1 second is sufficient, and so on. The other important matter is the accuracy of the gate control signal. The overall accuracy of the instrument is dependent upon the accuracy of the gate control signal. Typically, the time base generator uses some sort of a crystal oscillator followed by a chain of frequency dividers to generate the gate control signal. The divider chain inside the time base generator also provides the user the option to choose different values of gate control signals for a particular resolution. High-quality (more accurate) frequency counters employ a stabilization circuit within the time base generator circuit to minimize short- and long-term fluctuations in the time base signal.

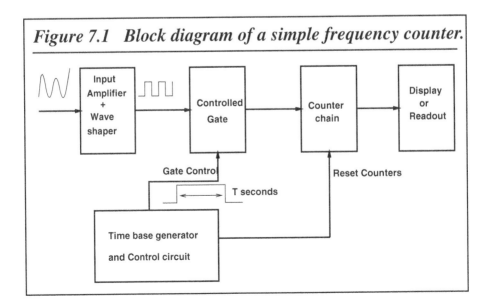

Figure 7.1 Block diagram of a simple frequency counter.

Besides generating the time base signal (i.e., the gate control signal), this part of the frequency counter also controls the chain of counters with their inputs connected to the controlled gate. At the beginning of a measurement cycle, the control circuit generates a signal to clear the contents of the counters and reset them to zero, then the gate control signal is generated. Because the counter chain has a zero value before the gate is enabled, the count in the chain of counters at the end of the gate control signal shows the number of pulses that it has accumulated in that period. From this information, you can easily calculate the frequency of the input signal. In simple circuits, this calculation involves mere shifting of the decimal point to an appropriate location.

The outputs of the chain of counters is connected to the display or a suitable read-out circuit. In stand-alone devices, this would be some sort of seven-segment LED or LCD display. For computer-controlled devices, this circuit would be composed of suitable latches and buffer ICs. The period counter works in a similar fashion to a frequency counter, as you can see from the block diagram in Figure 7.2. The fundamental difference between the frequency counter and the period counter is the nomenclature of the gate control signal. In the case of the frequency counter, the gate control signal is generated locally and is used to count the number of pulses during a given time period. In the case of the period counter, the time base generator generates a high-frequency signal (much higher than the frequency of the input signal whose period is to be measured). The gate control circuit is controlled by the input signal, which allows the counters to be clocked by the high-frequency signal from the time base generator. At the end of each period of the input signal, the count accumulated by the counters is proportional to the period of the input signal. Suppose the time base generator is set to a frequency 1MHz. If the input signal is 100Hz (i.e., a time period

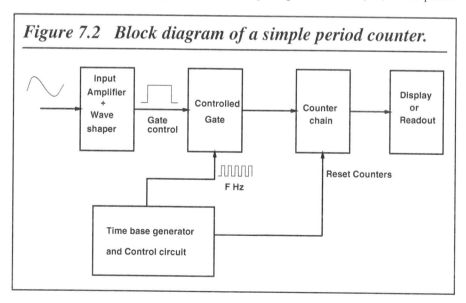

Figure 7.2   Block diagram of a simple period counter.

of 10ms), the count accumulated in the counters will be 10,000. Because the units of the time base signal is 1µs, the input period will be seen as 10,000µs.

For low-frequency signals, it is convenient to measure the period of the signal and to calculate the frequency from these figures rather than measure the frequency directly. At high frequencies, it is advantageous to measure the frequency and then interpret the period if required.

You can easily build either a stand-alone frequency/period counter with an integral display or a computer-controlled frequency/period counter. The advantage of the computer-controlled instrument is that the resultant data can be processed within the computer.

In the next section, I will describe a general-purpose, inexpensive frequency and period counter interface for the PC. My goal is to demonstrate the principles behind the operation of a parallel-port-controlled frequency and period counter If you adapt this design for your own uses, there are many areas in which you may wish to choose different components. I have used an inexpensive time base generator with a clock crystal of 32,768Hz. You may want to replace this with more accurate 1 or 10MHz crystal components. However, changing the crystal may require replacing some of the CMOS components, which may not work at the high 10MHz frequency. Similar components in the HCTTL family should be available.

My idea of using a low-frequency crystal and CMOS components was to keep the current consumption down. The circuit (as shown in the schematic in the next section) consumed a mere 5mA current. This tiny current requirement can even be met by squeezing current out of the PC's serial port, an idea extremely attractive and enterprising for portable applications.

Measuring Time Period and Frequency Using Discrete Components

The block diagram in Figure 7.3 shows the scheme for the PC parallel-adapter-controlled frequency and time period counter. The input signal passes through the block labeled Input Signal Conditioner (which, in the actual circuit, corresponds to a pair of diodes that protect the circuit from accidental overvoltages). The output of this block is fed to a digital multiplexer, which is also fed the output of the time base generator block. The time base generator produces two signals, with frequencies of 1 and 32,768Hz.

The digital multiplexer block has two channels, with inputs A1, A2 and B1, B2 and outputs Y1, Y2. The Y outputs get the signal on either pin A or B depending on the Mux control signal. When the Mux control signal is 0, Y1 = A1 and Y2 = A2. A1 is connected to the input signal and A2 is connected to the time base frequency of 32,768Hz. This is the period counter mode of operation. To begin a cycle of acquisition, the PC program resets the counters through the parallel adapter. The Y1 output of the multiplexer (which is the input signal) is divided by two, so the ON time of the resultant signal is equal to the period of the input signal.

Figure 7.4 shows the timing diagram for the instrument's period counter mode. The input pulse is converted into a square wave by the divider circuit. The high time of the resultant signal is equal to the period of the input signal. This signal acts as the gate control signal for the gate. The other signal to the gate is the output of the time base generator, which in our case this is a digital wave with a frequency of 32,768Hz (a period of about 30.5μs). The forth signal in the timing diagram is the output of the gate that goes to the counter and readout circuit.

A PC program controls the counter circuit by monitoring the gate control signal. At the time of an acquisition cycle, the program resets the counters and waits for one high period of the gate control signal to elapse. After the high period of the gate control signal elapses, the PC reads out the counters through the STATUS port input bits.

In the frequency counter mode of the circuit, the roles of the input signal and the time base signals are interchanged. The multiplexer is so arranged that now the time base generator signal of 1Hz is used as the gate control signal. This is shown in the

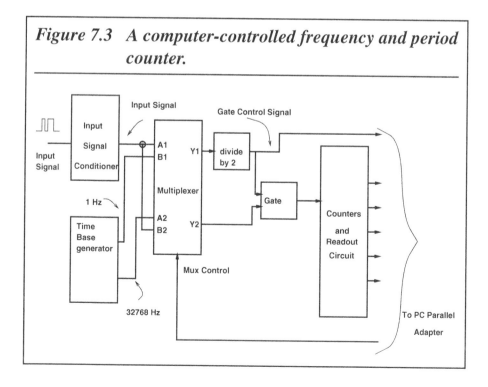

Figure 7.3 A computer-controlled frequency and period counter.

timing diagram of the frequency counting mode in Figure 7.5. The 1Hz signal at the output Y1 of the multiplexer is divided by the divider circuit to generate a high-time pulse of one second, as shown in the timing diagram. The other output of the multiplexer, Y2, is now the input signal of whatever frequency. The counters are now allowed to increment for a duration of one second by the gate control signal, during which time they accumulate a count equal to the frequency of the input signal. At the end of the one-second period, the PC again reads out the accumulated count through the STATUS port signals, as in the previous case.

Figure 7.6 shows the circuit schematic for the frequency and period counter. The input signal is connected through the 1Kohm resistor to the 1N4148 diode pair, which provide a nominal protection to the circuit against accidental overvoltages. This signal

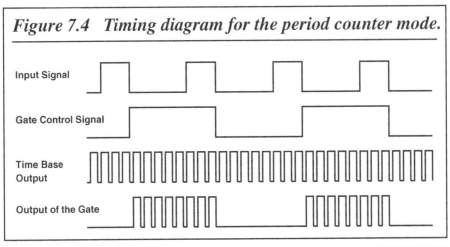

Figure 7.4   Timing diagram for the period counter mode.

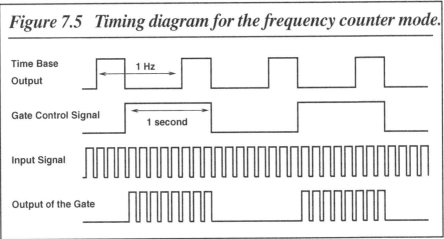

Figure 7.5   Timing diagram for the frequency counter mode.

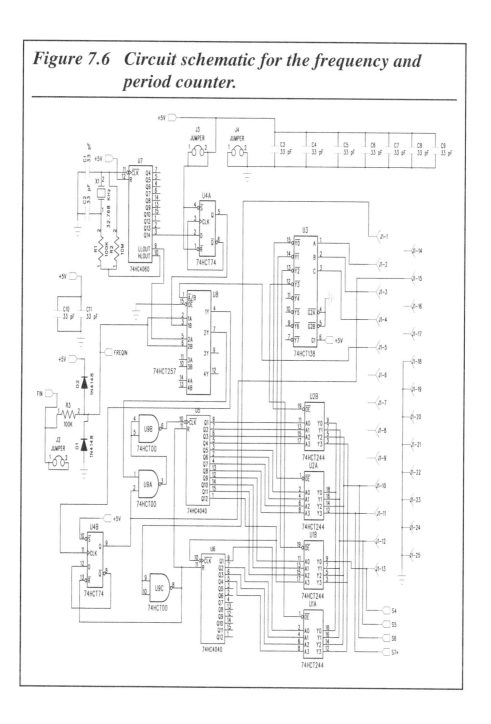

Figure 7.6 Circuit schematic for the frequency and period counter.

is then connected to the IC U8 (74HCT257), which is a digital multiplexer. IC 74HCT257 is a four-channel multiplexer, of which only two channels are used.

The input signal is connected to the A1 and B2 inputs of the two channels. The time base generator circuit is built around IC U7 (74HC4060), which is an oscillator-cum-binary counter IC. The IC provides a digital signal of frequency equal to the crystal, which in our case is the 32,768Hz crystal. The other output of the IC is a 0.5Hz signal, which is divided with the help of a D flip-flop to provide a 1Hz signal. The 32,768 and 1Hz signals are connected to the A2 and B1 inputs of the digital multiplexer.

One of the outputs of the multiplexer, Y1, is connected to another D flip-flop to provide a division of two. The output of the flip-flop is connected to the input of a NAND gate, U9A. The output of the flip-flop is also connected to the STATUS port signal S3. The gate is formed out of NAND gates U9A and 9B. The output of the gate is connected to the counter inputs. The other input to the gate is the output Y2 of the digital multiplexer.

The counters are formed out of inexpensive 74HC4040 binary counters. These are 12-bit counters, and you need two of them to provide a 16-bit counter — a 24-bit count option is also available. The counters and the divide-by-two flip-flop (U4B) can be reset under program control with the help of signal C0* of the CONTROL port. The outputs of the counters are connected to the tristate buffer IC (74HCT244) inputs. The buffers have their outputs shorted such that only four signals result. (You saw this same trick in a previous chapter.) Data flow through the tristate buffers is controlled by the outputs of the 3-to-8 decoder IC, U3 (74HCT138). The decoder IC is driven by the DATA port signals D0, D1, and D2. Signal D3 of the DATA port is connected to the mux control input of the digital multiplexer and is used to control the frequency/period counter mode of operation of the circuit.

The acquisition cycle first selects the correct mode of operation (frequency counter or period counter) through the D3 signal of the DATA port. Then the program resets the counters and the flip-flop (U4B) and starts monitoring the output of this flip-flop. It allows the output to go from logic 0 to logic 1 and back to logic 0. At this time, the program reads the counters by manipulating the D0, D1, and D2 signals.

For frequency counter mode, the count obtained by reading the binary counters is the frequency of the input signal with a resolution of 1Hz. For the time counter option, the resultant count must by normalized by multiplying it with the period of the time base generator circuit (32,768Hz).

An Astronomical Photometer Interface

Here is a variation of the frequency counter circuit that is very useful to an astronomer. In my laboratory, I needed a computer interface for a photometer (SSP3 by OPTEC) for photometric observations. The photometer produces low pulse width pulses. The frequency of these pulses is proportional to the incident light intensity. Photometric observations involve accumulating the pulses for a required period of time. This is easily achieved by gating a counter with a presettable timer. An 8053 programmable timer type device with three timers/counters is well suited for this application. An 8053 device can be interfaced to the PC via the ISA BUS. You can configure the 8053 to have one timer to gate the incoming pulses to the two other counters cascaded to form a 32 bit counter. The 8053 timer will allow a high-resolution gating period. In practice, however, such a wide selection of gating time is hardly used.

I, however, decided to use the printer adapter to connect the photometer. This choice was also dictated by the need to share the photometer between many computers without having to remove the computer interface from inside the computer. An added penalty for this approach could be an extra power supply for the circuit. However, the need for additional power was avoided in this case by using the RS-232 signal lines to power the frugal requirements of the interface circuit.

The block diagram in Figure 7.7 shows what is required of this circuit. The pulses from the photometer connect to the gate circuit. The gate is controlled by the output of a digital multiplexer. The digital multiplexer has eight inputs, which are outputs of the time base generator circuit. These signals are square waves of the required integration window period.

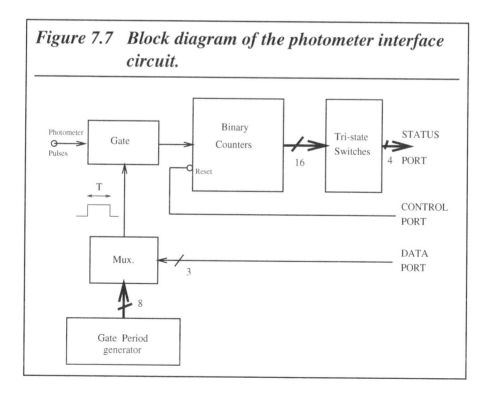

Figure 7.7   Block diagram of the photometer interface circuit.

One of these eight signals is selected by the multiplexer and acts as the gate control signal. The selection of the required period is done by the three DATA port signals. The gate output is connected to the chain of binary counters and the output of these counters drives the tristate buffer ICs connected in the usual fashion to the four STATUS port signals. At the beginning of each acquisition cycle, the counters are cleared by one of the CONTROL port signals.

Figure 7.8 shows the acquisition circuit schematic. Using CMOS-type components keeps the current consumption small; sufficiently small to be powered by the RS-232 signal lines. IC U7 (74HC4060) uses a 32,768Hz crystal to generate the gating signal of $^1/_8$, $^1/_4$, and $^1/_2$Hz. This signal is further divided by U8 (74HCT4024) to generate the rest of the five periods. These eight gating signals drive U4 (74HCT151), a 1-of-8 digital multiplexer. Three lower bits of the printer adapter DATA PORT are used as the select input of this multiplexer.

The output of the multiplexer drives U9-A (74HCT74) and a D type flip-flop in a divide-by-two mode to generate a signal with ON time from 0.125 to 16 seconds. The output of the flip-flop gates the input pulses through the AND gate (U10-A, B) to the counter chain. During each cycle of acquisition, the CONTROL PORT bit C0 is used to reset the flip-flop U9-A and the binary counters U5 and U6. A rising edge of the gate signal sets U9-A, enabling the incoming pulses to reach the binary counter chain U5 and U6, which are 12-bit binary counters. At the end of the gate period, the counters are frozen with the accumulated count. Of the twenty-four bits output from these counters, only sixteen (four nibbles) can be read by the program. The sixteen bits are multiplexed by U1 and U2 (74HCT244) to generate four bits. IC U3 (74HCT138) decides which one of the four nibbles is routed to the STATUS port. The output of the counters U5 and U6 is read by the program four bits at a time by reading the STATUS port and manipulating the four nibbles to form a 16-bit integer.

Figure 7.9 shows the power supply that is used to power the photometer interface circuit. Because the consumption of the circuit is small (less than 5mA), it can be easily powered by squeezing this current out of the RS-232 signal lines. The RS-232 lines are driven to +Vcc and passed through a 78L05 regulator to produce the +5V needed by the circuit. Note that a 78L05-type regulator is not ideally suited for such applications. This is because the regulator itself needs about 3mA of quiescent current. Another option for the power supply circuit is to use the 5V low-power zener IC (LM335). The zener IC requires a mere 400μA quiescent current and is more suitable than the hungry 78L05.

The photometer generating TTL level pulses is connected to the interface by a suitable cable. The control program (Listing 7.1) prompts the user to enter the required gate period and then starts acquiring data. The data is displayed on the screen as it is acquired. The program can be modified to store the numbers in a user-defined output file at the end of the acquisition.

Figure 7.8   Photometer interface circuit schematic.

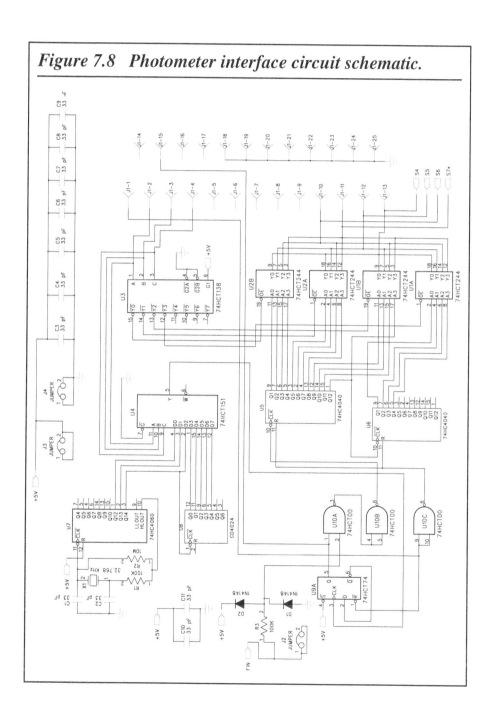

Figure 7.9 **A power supply derived from the RS-232 serial port signal pins can supply current up to 10mA, which is sufficient for the photometer circuit.**

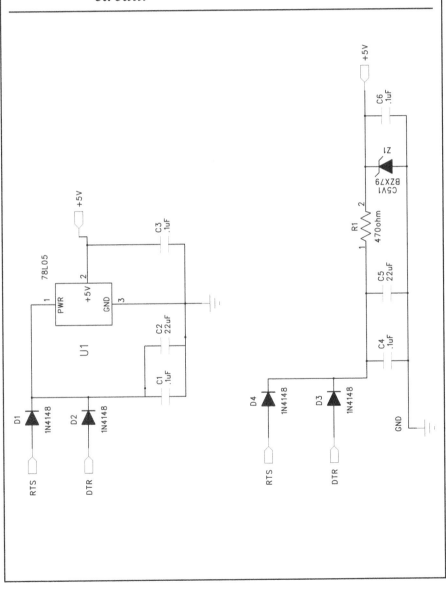

Listing 7.1   The photometer interface.

```
/*photo.c*/
/**********************************************************************/
/* Program to demonstrate the working of the Photometer interface */
/* through the parallel printer adapter. Connect the interface to */
/* LPT1 & COM1. Connect the photometer to the interface.          */
/**********************************************************************/
/*                 Compile with TurboC Ver. 2.0                    */
/**********************************************************************/

#include <stdio.h>
#include <dos.h>
#include <conio.h>
#include <process.h>
int modem_control_reg, dport_lpt1, cport_lpt1, sport_lpt1;

/* Declare the subroutines */
unsigned char get_integ_time(void);
unsigned int get_count(void);
unsigned char get_integ_time(void)
{
    int temp;
    clrscr();
    printf("Character     Integration time\n");
    printf("0             00.125 sec\n");
    printf("1             00.250 sec\n");
    printf("2             00.5   sec\n");
    printf("3             01.0   sec\n");
    printf("4             02.0   sec\n");
    printf("5             04.0   sec\n");
    printf("6             08.0   sec\n");
    printf("7             16.0   sec\n");

    temp=getchar();
    switch(temp)
      {
      case '0':    return 0;break;
      case '1':    return 1;break;
      case '2':    return 2;break;
      case '3':    return 3;break;
      case '4':    return 4;break;
      case '5':    return 5;break;
      case '6':    return 6;break;
      case '7':    return 7;break;
      default:     return 100;
      }
}
```

Listing 7.1 (continued)

```c
unsigned int get_count(void)
{
    unsigned char monitor, nib_0, nib_1, nib_2, nib_3;

    /* Reset the binary counters U6, U7 & the flipflop U4-A */
    /* The Control port bit C0 of the printer adapter is used */
    outportb(cport_lpt1,1);
    outportb(cport_lpt1,0);

    /* Monitor the o/p of flipflop U4-A */

    /* Wait in the loop till it is zero */
    monitor = 0;
    while(monitor == 0) monitor = inportb(sport_lpt1) & 0x08;

    /* Now the flipflop has enabled the counters */
    /* So, wait till the counting is over   */
    monitor = 0x08;
    while(monitor == 0x08) monitor = inportb(sport_lpt1) & 0x08;

    /* Read the four nibbles*/
    /* Nibble 3, the most significant nibble*/
    outportb(dport_lpt1,3);
    nib_3 = inportb(sport_lpt1);

    /* Nibble 2 */
    outportb(dport_lpt1,02);
    nib_2 = (inportb(sport_lpt1) >>4);

    /* Nibble 1 */
    outportb(dport_lpt1,1);
    nib_1 = inportb(sport_lpt1);

    /* Nibble 0, the least significant */
    outportb(dport_lpt1,0);
    nib_0 =(inportb(sport_lpt1) >> 4);

    /* put all the nibbles in one integer variable */

    nib_1 = nib_1 & 0x0f0;
    nib_1 = nib_1 | nib_0;
    nib_1 = nib_1 ^ 0x88;

    nib_3=nib_3 & 0x0f0;
    nib_3 = nib_3 | nib_2;
    nib_3 = nib_3 ^ 0x88;
```

Listing 7.1 (continued)

```c
   /* Return the unsigned integer */
   return (nib_3*256 + nib_1);
}

void main(void)
{
   unsigned char int_win;

   /* Sign ON */
   clrscr();
   printf("Photometer interface for the IBM PCs & compatibles, Version   1.0");
   printf("\nUses the parallel printer adapter for counting");
   printf("\n&the RS-232 port for the power supply");
   printf("\n\nBy: D.V.GADRE");

   /* Get COM1 register address */
   modem_control_reg = peek(0x40,0)+4;
   if(modem_control_reg == 4)
       {
       printf("\n\n\nCOM1 not available... aborting\n\n\n");
       exit(1);
       }
   printf("\n\n\nCOM1 address = %X",peek(0x40,0) );

   /*Get LPT1 port addresses */
   dport_lpt1 = peek(0x40,0x08);

   if(dport_lpt1 == 0)
       {
       printf("\n\n\nLPT1 not available... aborting\n\n\n");
       exit(1);
       }
   printf("\nLPT1 address = %X", dport_lpt1);
   cport_lpt1 = dport_lpt1 +2;      /* control port address */
   sport_lpt1 = dport_lpt1 + 1;     /* status port address */

   /* put power On on COM1 */
   outportb(modem_control_reg, 03);
   printf("\nPutting Power ON...\n\n\n");
   sleep(1);

   /* Get the integration time window */
   int_win = get_integ_time();

   while(int_win<0 || int_win >7) int_win = get_integ_time();
   clrscr();
```

Listing 7.1 (continued)

```
printf("Using integration window %d\n\n", int_win);
while(!kbhit())
    {
    /* set integration time */
    outportb(dport_lpt1, int_win);

    /* Get the accumulated count & print it on the screen */
    printf("%u ", get_count());
    }
}
```

## Chapter 8

Complete
Data Acquisition Systems

Now that I have described how to interface various digital components, it is time to build a complete interface system. A complete interface system offers the following.

- Analog input
- Analog output
- Digital input
- Digital output

The choice of the components for the analog input and output is determined by the required speed of conversion and the required resolution. This chapter describes designs that offer 8-bit and 12-bit resolution. Both solutions are general-purpose designs and can be adapted as required for special situations. You could also build similar systems using other components (such as timers, ADCs, and DACs).

Auto-Powered, 8-Bit ADC Interface

Before describing the two complete designs, I will begin with a very simple 8-bit, single-channel ADC that you can easily build with only a handful of components: two ICs, a couple of diodes, capacitors, and a zener. The circuit can fit on a small general-purpose PCB, which you can mount directly on the printer port connector using a mating connector. The power supply for the ADC is supplied by the RS-232 port of the PC.

This auto or "zero-powered" ADC circuit is built around the popular and inexpensive 8-bit ADC, the ADC0804. I described the functioning of this ADC in some detail in Chapter 6.

Figure 8.1 shows the schematic of an 8-bit ADC for the parallel adapter using only two ICs: an ADC0804 ADC and a 74HCT244 tristate buffer. Also shown is the power supply for the circuit. The circuit is set up such that the ADC conversion process is triggered by the C3* bit of the CONTROL port. The converted data is controlled by the C2 bit of the CONTROL port connected to the RD* pin of the ADC. After the ADC conversion is complete, the C2 bit is taken low and the conversion data appears on the output data lines of the ADC. Then the data is read into the PC with the help of the four STATUS port lines: S4, S5, S6, and S7*. The eight data output bits of the ADC are routed through the tristate buffer IC such that only four bits of the ADC data appear on the output lines of the tristate buffer at a time. The two buffer sections of the tristate IC are controlled by the C0* and the C1* bits of the CONTROL port. Initially, both lines are held high.

The conversion process of the ADC is triggered by taking the C3* line momentarily low and then high again. The end of conversion is monitored by the INTR output of the ADC, which is read through the S3 bit of the adapter's STATUS port. At the end of conversion, the C2 line is taken low. This makes the conversion data appear at the inputs of the buffer IC. Now C1* is taken low, and the STATUS port is read once; C1* is taken high again, and C0* is taken low. This makes the high nibble of the ADC data appear on the STATUS port lines. The STATUS port is read again, and the C0* signal is taken high again. The program then takes the two nibbles that were the result of reading the STATUS port twice and makes a byte out of them. This byte is the result of the ADC conversion. A fresh conversion can now begin.

The power to the circuit is taken out of the PC's COM1 serial RS-232 port. That is why the system is referred to as an auto-powered or zero-powered ADC system. The ADC requires a mere 1.5mA maximum current for operation.

Listing 8.1 is a program that operates the ADC system and checks if the ADC is connected. It then triggers conversion on the ADC, reads the ADC data, and displays the data on the PC screen.

It should be noted that, in the form shown in Figure 8.1, the ADC can only sample very low frequency input analog signals (not more than a few hundred Hertz) because the sample-and-hold amplifier has been omitted.

A Complete 8-Bit Interface Package

This section describes a circuit scheme to implement a package that offers analog I/O with a resolution of eight bits. The most important component of the circuit is the

Figure 8.1 Circuit schematic for a simple auto-powered 8-bit ADC.

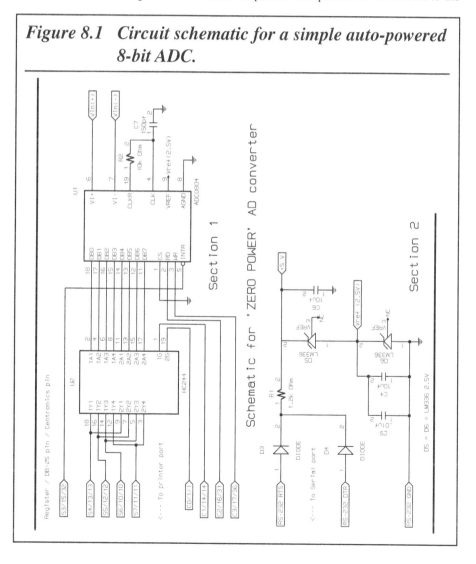

ADC interface. For this circuit, I have chosen a fast, 8-channel, 8-bit ADC — the MAX158 from Maxim. Besides the ADC, the circuit offers a single-channel 8-bit DAC, 8-bit output latch for digital output, and an 8-bit digital input buffer, read as two 4-bit nibbles.

The features of this package are:

- 8-channel, 8-bit ADC with a 2.8µs conversion time;
- 8-bit digital output to connect to an 8-bit DAC, such as a DAC08;
- eight bits of digital output; and
- eight bits of digital input.

The block diagram of the circuit is shown in Figure 8.2. The ADC has eight analog input channels. The output of the ADC conversion is available on the eight data output pins of the ADC. To select a particular channel for conversion, three of the eight channel address pins — A0, A1, and A2 — are available. The user puts a 3-bit number on these pins before initiating a conversion. After a conversion is initiated by the user, the ADC signals that the conversion is over on the INT* pin of the ADC.

Listing 8.1   Program to operate the ADC system.

```
/*zero_pwr.c*/
#include <stdio.h>
#include <dos.h>
#include <conio.h>
#include <process.h>
                                  /* C3* C2 C1* C0* */
#define RESET_VALUE          0X04  /* 0   1  0   0    */
#define START_ADC_CONV       0X0c  /* 1   1  0   0    */
#define READ_UPPER_NIBBLE    0X01  /* 0   0  0   1    */
#define    READ_LOWER_NIBBLE 0X02  /* 0   0  1   0    */

void main(void)
{
   int modem_control_reg, dport_lpt1, cport_lpt1, sport_lpt1;
   unsigned char adc_status, adc_val, upper_nib, lower_nib, intr_status;

   /* Sign ON */
   clrscr();
   printf("Zero Power (Well, almost!) ADC for the printer adapter, Version 1.0");
   printf("\nD.V.GADRE");

   /* Get COM1 register address */
   modem_control_reg = peek(0x40,0)+4;
   if(modem_control_reg == 4)
      {
      printf("\n\n\nCOM1 not available... aborting\n\n\n");
      exit(1);
      }
   printf("\n\n\nCOM1 address = %X",peek(0x40,0) );
```

Figure 8.2 shows how to interface a high-speed, 8-channel, 8-bit ADC to the printer port. The unused decoder outputs are used to connect other things, such as a digital output latch and a DAC, to make a complete analog–digital I/O system.

MAX158 Features and Mode of Operation

Figure 8.3 is a simplified timing diagram of the MAX158 ADC conversion and read-out process. To appreciate the diagram, you must understand the meaning of the various timing characteristics symbols (Table 8.1). The first signal in Figure 8.3 is CS*, the Chip Select signal. This is an active low signal. To initiate a conversion and to read

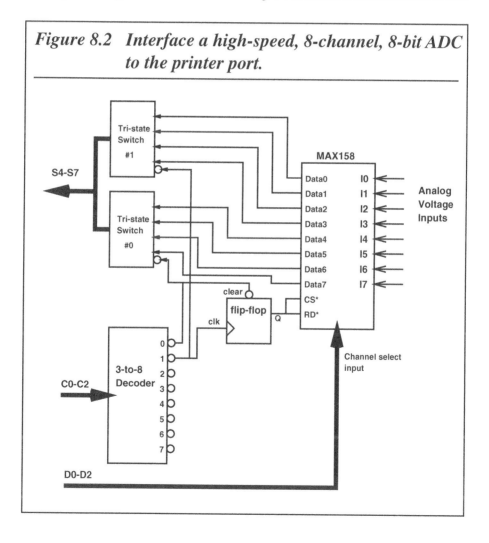

Figure 8.2   Interface a high-speed, 8-channel, 8-bit ADC to the printer port.

back the converted value, CS* must be held low. Next, RD*, the Read signal to the ADC, must be brought to logic 0. Tcss is the time difference between these two signals. From the timing characteristics table, the Tcss minimum accepted value is zero, which means that CS* and RD* can both be pulsed low at the same time. To initiate a conversion, the ADC needs a 3-bit value for the channel number on which the conversion will be started, as shown by the third signal on the timing diagram. The address

Listing 8.1 *(continued)*

```c
/*Get LPT1 port addresses */
dport_lpt1 = peek(0x40,0x08);
if(dport_lpt1 ==0)
    {
    printf("\n\n\nLPT1 not available... aborting\n\n\n");
    exit(1);
    }
printf("\nLPT1 address = %X", dport_lpt1);
cport_lpt1 = dport_lpt1 +2;    /* control port address */
sport_lpt1 = dport_lpt1 + 1;    /* status port address */

/* put power On on COM1 */
outportb(modem_control_reg, 03);
printf("\nPutting Power ON...");

/* check if ADC is connected & working*/
/*start ADC conversion & wait for 1 ms, this puts INTR to logic '0' */
/* reset the control port */
outportb(cport_lpt1, RESET_VALUE);
sleep(1);
/* start conversion */
outportb(cport_lpt1, START_ADC_CONV);
outportb(cport_lpt1, RESET_VALUE );
sleep(1);

/* hopefully the conversion is over, so read the INTR status */
/* if everything is OK, INTR should be '0' */
adc_status = inportb(sport_lpt1) & 0x08;
outportb(cport_lpt1, READ_LOWER_NIBBLE);
outportb(cport_lpt1, RESET_VALUE);

/* read the INTR status again */
/* if everything is OK, INTR should be '1' */
intr_status = inportb(sport_lpt1) & 0x08;
if( !( (adc_status == 0) && (intr_status == 0x08) )  )
    {
    printf("\n\n\nADC not connected... aborting\n\n\n");
    exit(1);
    }

/* acquire ADC sample */
while(!kbhit())
```

for the required channel must be set *Tas* seconds before the falling edge of the RD*
signal. *Tas* minimum is zero. The ADC now starts a conversion. After a time of *Tcrd*
seconds, which is the conversion time of the ADC, the ADC completes the conversion
and indicates that it is finished by taking the INT* signal low. INT* is the fourth signal shown in the timing diagram. Time *Tcrd* has a minimum of 1.6μs and a maximum
of 2.8μs. Thus the conversion time for the ADC lies between 1.6 and 2.8μs. After time
Tacc2 of the falling edge of the INT* signal, the converted data appears on the eight
data lines. Tacc2 is the data access time and the maximum value is 50ns. During the
conversion, the data lines are in a high-impedance state, as shown by the last signal in
the timing diagram.

Table 8.1 Timing characteristics symbols.

Symbol	Parameter	Min.	Max.	Units
Tcss	CS* to RD* setup time	0		ns
Tas	Multiplexer address setup time	0		ns
Tcrd	Conversion time	1.6	2.8	μs
Tacc2	Data access time		50	ns
Tinth	RD* to INT* delay		75	ns
Tcsh	CS* to RD* hold time	0		ns

Figure 8.3 A simplified timing diagram shows the various signals associated with the MAX158 ADC.

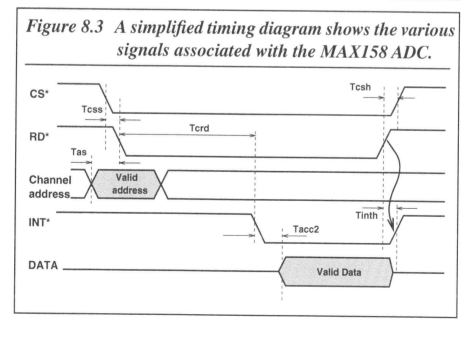

The converted data is available on the data lines until the RD* and the CS* signals are both low. If the RD* and CS* signals are taken high, it indicates that the data has been read. INT* then goes high within a maximum of 75ns after the RD* is taken high, and the data lines go into high-impedance state again. (The CS* signal can be taken high thereafter or even at the same time as the low-to-high transition of the RD* signal because *Tcsh* minimum time is 0ns.) A fresh conversion can now be initiated by pulsing CS* and RD* low again.

Now that I have describe the conversion and readout process for the MAX158, I will add the parallel port signals to the design. The block diagram (Figure 8.2) shows three CONTROL port signals, C0, C1, and C2, driving a 3-to-8 decoder. The decoder has eight, active low output signals numbered 0 to 7. Output 0 drives an active low clear signal of a D flip-flop. The Q output of this flip-flop provides the RD* and CS* signal to the ADC. Output 1 of the decoder drives the clock input of the flip-flop.

The decoder outputs also drive two 4-bit, tristate switches numbered 0 and 1. The four outputs of these switches are shorted to make four total outputs driving the STATUS input pins S4 to S7. The input to the tristate switches is the data output of the ADC.

The three channel address bits of the ADC are connected to the data output pins D0–D2 of the DATA port of the parallel port.

Listing 8.1 *(continued)*

```
{
outportb(cport_lpt1, RESET_VALUE);

start_conv:         outportb(cport_lpt1, START_ADC_CONV);
                    outportb(cport_lpt1, RESET_VALUE);
wait_for_conv:      adc_status = inportb(sport_lpt1) & 0x08;
                    while(adc_status)
                      {
                        adc_status = inportb(sport_lpt1) & 0x08;
                      }
read_upper_nibble:  outportb(cport_lpt1, READ_UPPER_NIBBLE);
                    upper_nib = inportb(sport_lpt1) & 0x0f0;
                    outportb(cport_lpt1, RESET_VALUE);
read_lower_nibble:  outportb(cport_lpt1, READ_LOWER_NIBBLE);
                    lower_nib = inportb(sport_lpt1) >> 4;
                    outportb(cport_lpt1, RESET_VALUE);
adc_val = (lower_nib | upper_nib) ^ 0x88;
delay(10);
printf("sample = %X ", adc_val);
  }
}
```

The conversion program on the PC sets the required address on the channel address pins and puts C0, C1, and C2 high. This enables output 7 of the decoder. To start a conversion, C0, C1, and C2 are made low, thus enabling output 0 of the decoder, which clears the flip-flop and Q goes to logic 0. From the previous discussion, you know that a high-to-low transition on the CS* and RD* signals will start a conversion. At the end of the conversion process, the data is available on the data output lines of the ADC, and in the present state of the logic, the lower nibble (Data0–Data3) is available on the STATUS port lines, which the conversion program reads. The program then changes C0, C1, and C2 such that decoder output 0 goes high and output 1 goes low. This does not change the flip-flop output and CS* and RD* are low. However, because decoder output 1 is low, the tristate switch number 1 is enabled and the higher nibble (Data4–Data7) of the ADC output data is routed to the STATUS port lines. The program then reads this nibble. By shifting and ANDing the two nibbles it has read, a complete byte is created, which is the ADC conversion result for the selected channel.

Now, the program changes C0, C1, and C2 such that decoder output 7 goes low and output 1 goes high. This low-to-high transition on output 1 clocks the flip-flop and its output Q goes high, thereby disabling the RD* and CS* signal of the ADC.

Circuit Schematic for the 8-Bit Package

The schematic diagram in Figure 8.4 shows the complete circuit diagram of the ADC interface, as well the digital output and digital input sections. The signals from the parallel port appear on the J1 DB-25 pin connector. These signals are:

- D0–D7, the DATA port output pins;
- C0*, C1*, C2, and C3*, the CONTROL port pins;
- S4, S5, S6, and S7*, the STATUS port pins;
- ground.

The CONTROL port signals C0*, C1*, and C2 drive the A, B, C inputs of IC 74HCT138 (U3), a 3-to-8 decoder. C3* is unused in this system. The decoder outputs are Y0 to Y7 and are labeled SEL_0 to SEL_7. SEL_0 is connected to pin 1 (clear input) of IC 74HCT74 (U4-A), a D type flip-flop. SEL_1 drives pin 3 (clock input) of the flip-flop. SEL_0 and SEL_1 also connect to pin 1 and 19 of IC 74HCT244 (U3), an octal tristate buffer. It is important to note here that only one of the eight outputs of the decoder can be low and the rest will be high.

The inputs of the tristate buffer (U3) are connected to the data outputs of the ADC DB0–DB7. The tristate switch IC has two sections. Section 1 has inputs 1As and outputs 1Ys; Section 2 has inputs 2As and outputs 2Ys. The value at the input is transmitted to the corresponding outputs, if the section enabling pin 1 and 19, respectively, is low, the section enable pin is active low. Thus if pin 1 is low, the inputs at section 1As

Figure 8.4 **Circuit schematic for a complete 8-bit analog and digital input and output.**

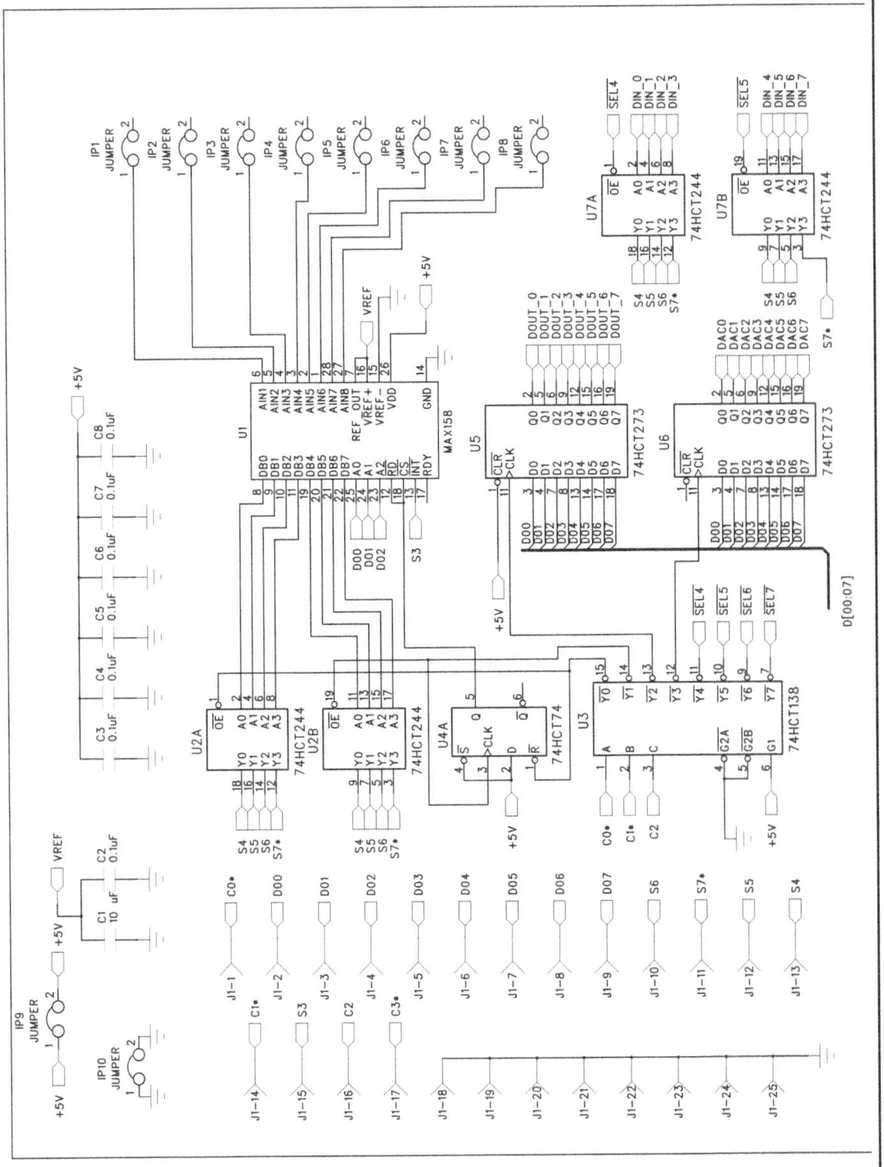

are output to the corresponding outputs 1Ys. If pin 1 is high, the outputs 1Ys go into a high-impedance state. In the circuit, pins 1 and 19 of the tristate buffer (U2) are driven by the decoder outputs, which means that only one of the two sections of the buffer will be enabled. The outputs of the other section will be in a high-impedance state.

Each output of two sections of the tristate buffer is shorted to the corresponding pin of the other section and then connected to the STATUS port pins. In this configuration, the STATUS port pins will read either DB0–DB3 or DB4–DB7 outputs of the ADC, depending on whether Section 1 or Section 2 of the buffer IC is enabled. If both the sections are disabled, the STATUS port pins will not be able to read any data from the ADC.

The ADC channel address pins A0, A1, and A2 are connected to D0, D1, and D2 pins of the DATA port. RD* and CS* signals are shorted and driven by the Q output of the flip-flop (IC U4-A). The end of conversion signal from the ADC, INT*, is connected to the STATUS port signal S3. The user program monitors the state of this pin before reading out the ADC data. The inputs of the ADC are labeled VIN_1 to VIN_8. The input range of the analog voltage that can be applied is determined by the voltage on signals Vref+ and Vref– of the ADC. I have connected Vref+ to the REF OUT signal and Vref– to ground. Nominally, the REF OUT signal of the ADC provides a stable +2.5V. So the range of the ADC input signals (VIN_1 to VIN_8) is 0–2.5V. Capacitors C1 and C2 provide noise filtering for the REF OUT signal and C3 and C4 filter the +5V supply to the ADC.

ICs U5 and U6 are 8-bit registers. The clock inputs of these ICs are driven by SEL_2 and SEL_3 signals of the decoder. The inputs of the ICs are driven by the DATA port output signals D0–D7. The outputs of these ICs are D_OUT0 to D_OUT7 and DAC0 to DAC7, respectively. D_OUT0 to D_OUT7 outputs provide the user with eight bits of digital outputs and can be used to drive relays, LEDs, etc. DAC0 to DAC7 signal outputs are supposed to be inputs to a relevant 8-bit DAC. If required, the user can use these outputs as ordinary digital output pins.

U7 is also an 8-bit tristate buffer IC. The inputs are labeled D_IN0 to D_IN7 and provide the user with eight bits of digital input. These inputs can be used to read digital signals from switches, sensors (with digital output levels), etc. The two sections of the buffer are enabled by signals from the SEL_4 and SEL_5 outputs of the decoder, respectively. Decoder outputs SEL_6 and SEL_7 are unused at the moment — you can use them for any for other tasks specific to your own design.

Listing 8.2 (`dio.c`) shows code to interface the various outputs and inputs of the system to the PC (i.e., how to read the ADC, how to read digital input bits, etc.).

A 12-Bit ADC/DAC Interface

The previous section described a complete 8-bit interface system. 8-bit systems are useful for a variety of routine data acquisition applications. However, in many applications, 8-bit resolution may be insufficient. How would you know if your resolution is insufficient? Say you want to read the ambient temperature, which is expected to lie between 0 and 50°C, and we want to resolve to $^1/_{25}$ of a degree. The resolution of the encoding device (i.e., the ADC) should be better than 1 part in 50 × 25 = 1,250. So clearly, an 8-bit ADC, which can resolve 1 part in 256, is not suitable. Even a 10-bit ADC (which can resolve 1 part in 1,024) would be unsuitable. A 12-bit ADC, which can resolve 1 part in 4,096, would be fine. Similarly, for different occasions, one would have to decide if the ADC (and similarly DAC) resolutions are meeting the requirement.

In this section I describe a 12-bit ADC system that is built around the popular MAX186 ADC from Maxim. The system offers 8-bit DAC using the MAX521 DAC from Maxim. Both devices are serial input and output devices. Connecting them to the

---

Listing 8.2   *Driver software for the 8-bit interface package.*

```c
/*dio.c*/
/*8-bit analog and digital I/O program*/
/*uses MAX158 ADC*/
#include <stdio.h>
#include <dos.h>
#include <conio.h>
#include <process.h>

/* used for the CONTROL port with C3* held high*/
/*C2, C1 and C0* are used as the inputs of a 3-8 decoder*/
                     /* C3*  C2   C1*  C0* */
#define LLL 3        /*  0    0    1    1   */
#define LLH 2        /*  0    0    1    0   */
#define LHL 1        /*  0    0    0    1   */
#define LHH 0        /*  0    0    0    0   */
#define HLL 7        /*  0    1    1    1   */
#define HLH 6        /*  0    1    1    0   */
#define HHL 5        /*  0    1    0    1   */
#define HHH 4        /*  0    1    0    0   */

#define WAIT_TIME 100000

unsigned int dport, cport, sport;

int chk_adc(void)
{
    unsigned char int_stat;
    /* check if ADC is connected & working*/
    /*put RD* and CS* to high. INT* should be high*/
    outportb(cport, LLH);
    outportb(cport, HHH);
    delay(10);   int_stat=inportb(sport);
    int_stat=int_stat & 0x08;
    /*printf("\nStat=%x", int_stat);*/
    if(int_stat != 0x08) return 0; /* ADC is not connected*/
```

Listing 8.2 *(continued)*

```
   outportb(cport, LLL);              /*trigger ADC conversion*/
   outportb(cport, HHH);
   delay(10);
   int stat=inportb(sport);
   int_stat=int_stat & 0x08;
   /*printf("\nStat=%x", int_stat);*/
   if(int_stat != 0) return 0;

   outportb(cport, LLH);
   /*else just complete the readout and return success*/
   outportb(cport, HHH);
   return 1;
}

int adc_convert(unsigned char chan_num, unsigned char *value_ptr)
{
   unsigned char low_nib, high_nib, stat;
   long timeout=0;

   timeout=0;
   chan_num=chan_num & 0x07;
   outportb(dport, chan_num);
   outportb(cport, LLL);              /*trigger ADC conversion*/
   stat= ( inportb(sport) & 0x08);    /*wait till conversion over*/
   while (stat == 0x08)
      {
      if(timeout > WAIT_TIME)return 0;   /*return for timeout*/
      stat= ( inportb(sport) & 0x08);
      timeout++;
      }

   low_nib=inportb(sport);
   outportb(cport, LLH);
   high_nib=inportb(sport);
   outportb(cport, HHH);

   *value_ptr = ( (low_nib >> 4) & 0x0f) | (high_nib & 0xf0);
   *value_ptr= *value_ptr ^ 0x88;
   return 1;
}

void read_digital_ip(unsigned char *digital_ip)
{
   unsigned char low, high;
   outportb(cport, HLL);
   low=inportb(sport);
   outportb(cport, HLH);
   high=inportb(sport);
   outportb(cport, HHH);

   *digital_ip= ( (low>>4) & 0x0f) | (high & 0xf0);
   *digital_ip=*digital_ip ^ 0x88;
   return;
}
```

Listing 8.2 (continued)

```c
void dout_port2(unsigned char value)
{
   outportb(dport, value);
   outportb(cport, LHL);
   outportb(cport, HHH);
   return;
}

void dac_out(unsigned char value)
{
   outportb(dport, value);
   outportb(cport, LHH);
   outportb(cport, HHH);
   return;
}

main()
{
   unsigned char adc_status, adc_val, upper_nib, lower_nib, intr_status;
   unsigned char result, channel=0;   float voltage;

   /* Sign ON */
   clrscr();
   printf("Fast, 8-bit, 8 channel ADC interface the printer adapter");
   printf("\nD.V.GADRE");

   /*Get LPT1 port addresses */
   dport = peek(0x40,0x08);
   if(dport ==0)
      {
      printf("\n\n\nLPT! not available... aborting\n\n\n");
      exit(1);
      }
   printf("\nLPT1 address = %X", dport);
   cport = dport +2;     /* control port address */
   sport = dport + 1;    /* status port address */

   outportb(cport, 0x04); /*Init the control port to all 1's*/
   outportb(cport, 0x02); /*pulse Y1 low of the decoder*/
   outportb(cport, 0x04);

   if( chk_adc() == 0)
      {
      printf("\n\n\nADC interface not connected or not powered... aborting\n\n\n");
      exit(1);
      }
```

parallel port is easy and uses only a few of the port signals, leaving quite a few of the port bits to the user for digital I/O.

The features of this system are:

- eight channels of 12-bit ADC inputs with a range of 0–4.095V;
- eight channels of 8 bit DAC outputs with a range of 0–5V;
- 12 digital lines for input and output.

Listing 8.2 (continued)

```
/* acquire ADC sample */
while(!kbhit())
    {
    if(adc_convert(channel, &result) == 0)
        {
        printf("\nError in ADC Conversion, channel %d. Aborting..", channel);
        exit(1);
        }

    printf("\nChannel %X, Voltage = %1.2f Volts", channel,
            0.01 * (float)result);
    sleep(1);
    }

/*Now reading data from the 8 bit digital i/p channel*/
while(!kbhit())
    {       read_digital_ip(&result);
    printf("\nDigital I/P = %X (hex)", result);
    sleep(1);
    }

/*Now writing data to the 8 bit digital o/p channel 2*/
while(!kbhit())
    {
    dout_port2(result++);
    sleep(1);
    }

/*Now writing data to the 8 bit digital o/p connected to a DAC*/
while(!kbhit())
    {
    dac_out(result++);
    sleep(1);
    }
}
```

Before I describe the circuit schematic, it is very important to understand the features and modes of operation of the ADC and DAC ICs. An understanding of the ADC and DAC is useful not only for understanding the circuit schematic and the driver routines, but also for altering the circuit schematic if you wish to configure the ADC/DAC system in some other way.

MAX186 ADC Features

MAX186 is a complete ADC system, combining an 8-channel analog multiplexer, a sample-and-hold amplifier, a serial data transfer interface, a voltage reference, and a 12-bit resolution successive approximation converter. All these features are packed into a 20-pin DIP package (other packaging styles are also offered). MAX186 consumes extremely low power and offers power-down modes and high conversion rates. The power-down modes can be invoked in software as well as hardware. The IC can operate from a single +5V as well as from a ±5V power supply. The analog inputs to

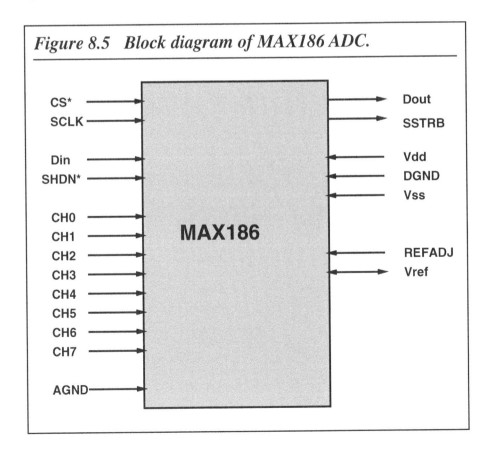

Figure 8.5   Block diagram of MAX186 ADC.

the ADC can be configured via software to accept either unipolar or bipolar voltages. The inputs can also be configured to operate as single-ended inputs or differential inputs. The ADC has an internal voltage reference source of 4.096V, but the user can choose not to use this reference and supply an external voltage between +2.50 and +5.0V. This gives the user the advantage of adjusting the span of the ADC according to need (e.g., if the input analog voltage is expected to be in the range of 0 to +3.0V, choosing a reference voltage of 3.0V will provide the user the entire ADC input range with a better resolution).

The MAX186 is an extremely fast device. It can convert at up to 133,000 samples/s at the fastest serial clock frequency. This ADC is best suited for devices with an

Table 8.2   MAX186 ADC signals.

Signal Name	Function
CS*	Active low chip select input.
SCLK	Serial clock input. Clocks data in and out of the ADC. In the external clock mode, the duty cycle must be 45–55%.
Din	Serial data input. Data is clocked at the rising edge of SCLK.
SHDN*	Three-level shutdown input. A low input puts the ADC in low power mode and conversions are stopped. A high input puts the reference buffer amplifier in internal compensation mode. A floating input puts it in external compensation mode.
CH0–CH7	Analog inputs.
AGND	Analog ground and input for single-ended conversions.
Dout	Serial data output. Data is clocked out at the falling edge of SCLK.
SSTRB	Serial strobe output. In external clock mode, it pulses high for one clock period before the MSB decision.
DGND	Digital ground.
Vdd	Positive supply voltage. +5V ±5%.
Vss	Negative supply voltage. –5V ±5% or AGND.
REFADJ	Input to the reference buffer amplifier.
Vref	Reference voltage for AD conversion. Also output of the reference buffer amplifier (+4.096V). Also, input for an external precision reference voltage source.

extremely fast serial port (e.g., some DSPs and microcontrollers). MAX186 is also an inexpensive device that is easily adapted to the parallel port, even though the parallel port is suited for reading parallel data. The trick is to convert the parallel port into a program-controlled serial device capable of shifting digital data in and out. The added penalty is the reduced conversion speeds. With contemporary PCs, the conversion rate using the parallel port is a little over 5,000 samples/s. However, the advantages of using the MAX186 (small size, very low power consumption, single supply operation) are too good to lose in favor of more exotic, parallel ADCs.

Figure 8.5 shows the block diagram of the MAX186 and its various associated signals. Table 8.2 identifies the function of each signal.

MAX186 Conversion and Readout Process

To initiate a conversion, you must supply the MAX186 with a control byte. The control byte is input into the ADC through the Din signal input. To clock the control byte, either an internally or externally generated clock signal (on the SCLK pin) could be used. To keep the hardware small and simple, it is necessary to use the external clock mode. The format of the control byte is shown in Figure 8.6.

To clock the control byte into the MAX186, pull the CS* pin low. A rising edge on SCLK clocks a bit into Din. The control byte format requires that the first bit to be

---

Figure 8.6   MAX186 control byte format.

MSB							LSB
START	SEL2	SEL1	SEL0	UNI/BP*	SGL/DF*	PD1	PD0

START: The first logic '1' bit after CS* goes low defines the start of the Control byte

SEL2, SEL1, SEL0: These 3 bits select which of the 8 channels will be used for conversion

UNI/BP*: 1=Unipolar; input can range between 0 to +Vref;
0=Bipolar; input can range between +Vref/2 to -Vref/2

SGL/DF*: 1=single ended; 0=Differential

PD1, PD0: Defines clock & power down modes.
0   0 : Full power down mode
0   1 : Fast power down mode
1   0 : Internal clock mode
1   1 : External clock mode

shifted in should be 1. This defines the beginning of the control byte. Until this start bit is clocked in, any number of 0s can be clocked in by the SCLK signal without any effect. The control byte must be 1XXXXXX11 binary. Xs denote the required channel and conversion mode. The data is shown as most significant bit ... least significant bit from left to right. The two least significant bits are 11 (to select the external clock mode option).The control byte value for starting a conversion on channel 0 in unipolar, single-ended conversion mode using the external clock is 10000011 binary or 83h.

Figure 8.7 shows the conversion and readout process on ADC channel 0. The timing diagram shows five traces, namely CS*, the chip select signal; SCLK, the serial clock required for programming the ADC and the subsequent readout; Din, which carries the programming information (the control byte); SSTRB, which the ADC generates to indicate the beginning of the readout process; and Dout, the actual data output from the ADC. The data on signal Din is clocked into the ADC at the rising edge of the SCLK signal. The first bit clocked in is D7. To begin the conversion, D7 needs to be set to 1, as can also be seen from the value of the control byte calculated earlier in this section. So, Din is set to 1 and the first SCLK rising edge is applied to the ADC. The SCLK is then taken low. Thereafter, Din is set to each of the subsequent bits of the control byte before applying SCLK. At the end of eight SCLK pulses, the

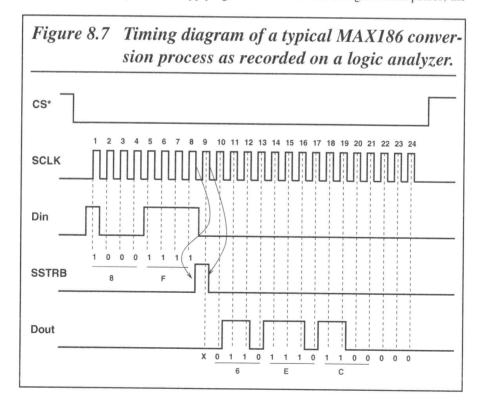

Figure 8.7 Timing diagram of a typical MAX186 conversion process as recorded on a logic analyzer.

Din bit is not required and is set to 0. At the falling edge of the eighth SCLK pulse, the ADC sets the SSTRB bit to 1. At the falling edge of the ninth SCLK bit, SSTRB is taken to 0.

At the rising edge of the ninth SCLK signal, the ADC outputs data on the Dout signal, one bit for each of the next 15 rising edges of the SCLK signal. The data on the ninth pulse is 0 and the actual conversion result is effective after the 10th rising edge to the 21st rising edge. Thereafter, for the next three edges, the ADC outputs 0s. To initiate a conversion and read out the result, a simple microprocessor circuit with minimal parts would need three output bits and one input bit. The output bits would be needed to generate the Din and SCLK signal and the input bit to read the Dout signal from the ADC.

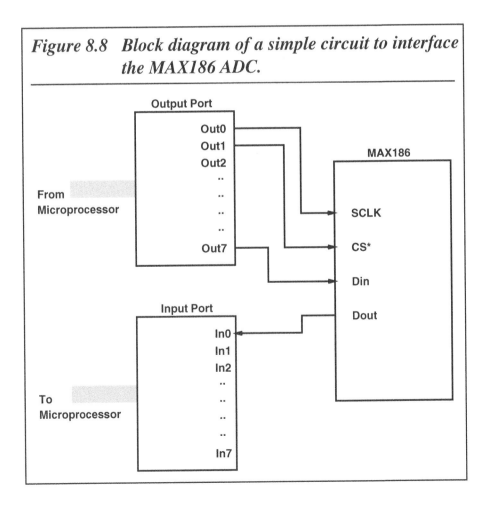

Figure 8.8 *Block diagram of a simple circuit to interface the MAX186 ADC.*

The diagram in Figure 8.8 shows an output port with output signals Out0–Out7 and an input port with signals In0–In7. The Din, SCLK, and CS* signals are driven by the output port signals and the Dout signal is connected to one of the bits of the input port. There is no particular reason to connect the signals in the order that they are shown, but as you will see, it makes programming a bit easier if Din and Dout are connected to an MSB and LSB, respectively, of the ports.

The input to the output port and the output of the input port are connected to the data bus of the microprocessor. The microprocessor can read the Dout signal by reading the port bit on the input port and can control the ADC signals CS*, SCLK, and Din by writing to the output port.

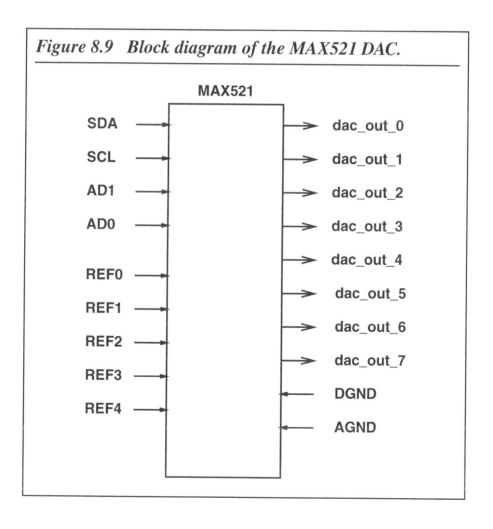

Figure 8.9   Block diagram of the MAX521 DAC.

The program running on the microprocessor begins by setting CS* low and then outputting the control byte on the Din signal, synchronized with the rising edge of the SCLK signal. After generating eight SCLK clock pulses, it starts reading the Dout pin of the ADC through the input port after each rising edge of the SCLK clock pulse for the next 16 SCLK clock pulses. The program then reconstructs the 12-bit result data from the 16 Dout values that it has previously read. After this, the microprocessor can initiate a new conversion and readout process.

The input and output signals that are needed by the ADC are generated by the DATA port, CONTROL port, and STATUS port bits.

MAX521 Features

This application uses the MAX521 to provide the user with eight channels of 8-bit DAC. Because this is a 12-bit system, you may be wondering why I have chosen an 8-bit part for the DAC. The answer is that, although the MAX521 is an 8-bit DAC system, its digital interface is common to many of the 12-bit DACs manufactured by Maxim, and you can easily replace the MAX521 with a suitable device and some modification to the driver software. The principle of connecting a 12-bit DAC is the same.

MAX521 is a voltage-out DAC and has a simple two-wire digital interface. These two wires can be connected to more MAX521s (a total of up to four). The IC operates from a single +5V supply. Even with a +5V supply, the outputs of the DACs can swing from 0 to +5V. The IC has five reference voltage inputs that have a range that can be set to anywhere between 0 and +5V. Figure 8.9 shows the block diagram of the MAX521. Table 8.3 lists the MAX521 DAC signals.

The MAX521 has five reference inputs. The first four DACs each have independent reference inputs and the last four share a common reference voltage input. The digital interface allows the IC to communicate to the host at a maximum of 400Kbps. The input of the DACs has a dual data buffer. One of the buffer outputs drives the DACs while the other can be loaded with a new input. All the DACs can be set to a new value independently or simultaneously. The IC can also be programmed to a low-power mode, during which the supply current is reduced to 4µA only. The power-on reset circuit inside the IC sets all the DAC outputs to 0V when power is initially applied.

The output of an 8-bit DAC is

$$V_{ref} = +V_{ref}(input/256)$$

where *input* is an 8-bit number and V_{ref} is the reference voltage for the channel.

Data Transfer to a MAX521

The operation of the MAX521 is slightly more complex than the operation of the MAX186. The MAX521 uses a simple two-wire interface. Up to four MAX521s can be connected to one set of these two-wire interfaces. This means that a host system with two output lines can be used to program up to 32 DACs! To send commands and data to the MAX521, the host sends logic sequences on the SDA and SCL lines. Otherwise, these lines are held to 1. The two-wire interface of the MAX521 is compatible with the I^{2}C interface. To maintain compatibility with I^{2}C, external pull-up resistors on the SDA and SCL lines are required.

MAX521 is a receive-only device, so it cannot transmit data. The host only needs two output signal lines for SDA and SCL signals. The SCL clock frequency is limited to 400KHz. The host starts communication by first sending the address of the device followed by the rest of the information, which could be a command byte or command byte and data byte pair. Each such transmission begins with a START condition, as

Table 8.3 Signal description of the MAX521 DAC.

Signal Name	Function
OUT0	DAC0 voltage output
OUT1	DAC1 voltage output
OUT2	DAC2 voltage output
OUT3	DAC3 voltage output
OUT4	DAC4 voltage output
OUT5	DAC5 voltage output
OUT6	DAC6 voltage output
OUT7	DAC7 voltage output
REF0	Reference voltage input for DAC0
REF1	Reference voltage input for DAC1
REF2	Reference voltage input for DAC2
REF3	Reference voltage input for DAC3
REF4	Reference voltage input for DACs 4, 5, 6, and 7
SCL	Serial clock input
SDA	Serial data input
AD0	Address input 0. Sets IC's slave address
AD1	Address input 1. Sets IC's slave address
Vdd	Power supply, +5V
DGND	Digital ground
AGND	Analog ground

shown in the timing diagram in Figure 8.10, followed by the device address (called the slave address) and command byte/data byte pairs or a command byte alone. The end of transmission is signaled by the STOP condition on the SDA and SCL lines.

The SDA signal is allowed to change only when the SCL signal is low, except during the START and STOP conditions. For the START condition, the SDA signal makes a high-to-low transition while the SCL signal is high. Data to the MAX521 is transmitted in 8-bit packets (which could be an address byte, the command byte, or the data byte) and it needs nine clock pulses on the SCL signal line. During the ninth SCL pulse, the SDA line is held low, as shown in the timing diagram. The STOP condition is signaled by a low-to-high transition on the SDA signal line when the SCL signal is held high.

The address and command bytes transfer important information to the MAX521. The address byte is needed to select one of a maximum of four devices that could be connected to the SDA–SCL signal lines. After the host begins communication with the START condition, all slave devices on the bus (here the bus is referred to the SDA and SCL signal lines) start listening. The first information byte is the address byte. The slave devices compare the address bits AD0 and AD1 with the AD0 and AD1 pin condition on the IC. In case a match occurs, the subsequent transmission is for that slave device. The next transmission is either a command byte or a command byte/data byte pair. In either case, the data byte, if present, follows the command byte, as shown in Figure 8.11. Table 8.4 shows the bit sequence of the command byte and the function of each bit.

All the possible combinations of address byte, command byte, and data byte for the MAX521 are:

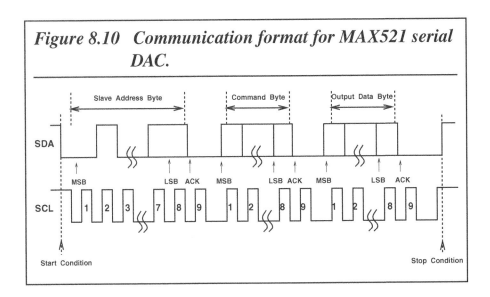

Figure 8.10   Communication format for MAX521 serial DAC.

1. START condition, slave address byte, command byte/output data byte pair, STOP condition.

2. START condition, slave address byte, command byte, STOP condition.

3. START condition, slave address byte, multiple command byte/output data byte pairs, STOP condition.

Figure 8.12 shows how to connect up to four MAX521s on a single bus from the host. The four devices are distinguished by the different addresses set on the AD0 and AD1 lines. Each of the MAX521s compares these bits with the address bits in the address byte transmission from the host.

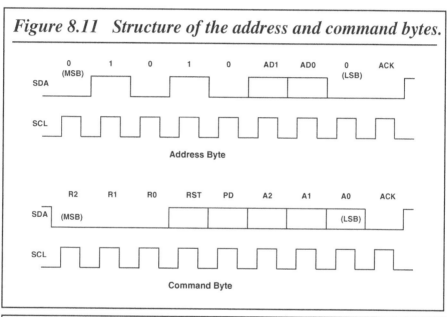

Figure 8.11   Structure of the address and command bytes.

Table 8.4   Bits of the command byte for the MAX521.

Bit Name	Function
R2, R1, R0	Reserved bits. Set to 0.
RST	RESET bit. A 1 on this bit resets all DAC registers.
PD	Power-down bit. A 1 on this bit puts the MAX521 in a power-down mode; a 0 returns the MAX521 to normal state.
A2, A1, A0	Address bits. defines address of the DAC to which the subsequent data byte will be addressed.
ACK	Acknowledgment bit. set to 0.

Circuit Schematic for the 12-Bit ADC/DAC System

Figure 8.13 shows the circuit schematic for the 12-bit MAX186 and 8-bit MAX521 DAC interface to the parallel port. Because the interface for the two devices is serial, the number of I/O lines required to connect the parallel port to the ADC and the DAC are very few.

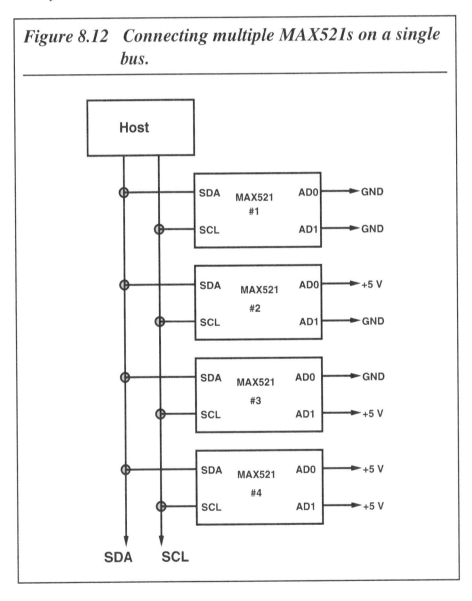

Figure 8.12   Connecting multiple MAX521s on a single bus.

The interface uses D7, D6, D5, and D0 bits of the DATA port; S7 of the STATUS port; and C0 of the CONTROL port. All the components and the 25-pin D male connector can be put on a small PCB the size of a small dongle, such that this attachment can be directly connected to the 25-pin D female connector of the PC parallel port.

Listing 8.3 is the driver software for the 12-bit ADC system.

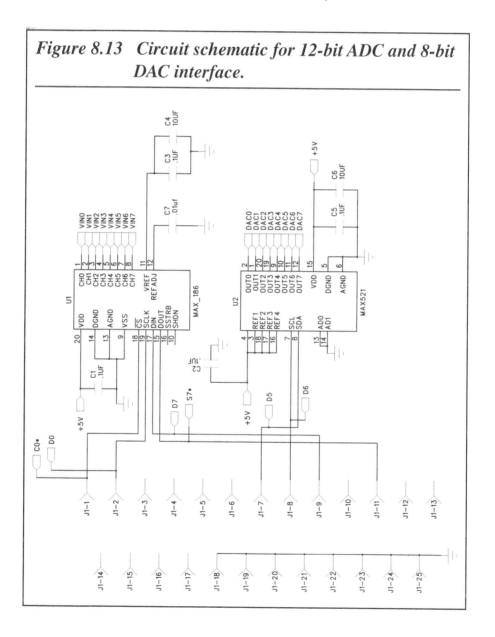

Figure 8.13   Circuit schematic for 12-bit ADC and 8-bit DAC interface.

Listing 8.3   Driver software for 12-bit ADC system.

```
/*ana_io.c*/

/*12bit, 8 channel ADC and 8 bit 8 channel DAC interface
Uses MAX186 and MAX521
Dhananjay V. Gadre
*/

#include<stdio.h>
#include<dos.h>
#include<time.h>
#include <conio.h>
#include <process.h>

/* D5 is SDA = data; D6 is SCL=clock */
#define H_H 0xff
#define H_L 0xdf
#define L_H 0xbf
#define L_L 0x9f

#define TRUE 1
#define FALSE 0

/*ADC control bytes for the 8 channels*/
unsigned char adc_con_byte[8];

/* Global variable has address of DATA port of LPT1 */
unsigned int dport_lpt1;                          /* data port address*/
unsigned char dport_value, cport_value; /*data and control port status*/

/*Check if LPT1 is present*/
int chk_lpt(void);

/* This generates the SLAVE address for the MAX-521 */
void set_up(void);

/*MAX521 start sequence*/
void st_seq(void);

/*MAX521 end sequence*/
void end_seq(void);

/* sets DAC address & the corresponding value */
void out_to_dac(unsigned int dac_number, unsigned int dac_value);

/*check if ADC is connected*/
int chk_adc(void);
```

Listing 8.3 (continued)

```c
/*start ADC conversion and get the result*/
int read_adc(int channel_number);

void st_seq(void)
{
    unsigned int tempa;
    tempa=inportb(dport_lpt1) & 0x9f;
    tempa=tempa | (0x60 & H_H);

    /* Generate start sequence for MAX-521*/
    outportb(dport_lpt1, tempa);
    tempa= tempa & 0x9f;
    tempa= tempa | (H_L & 0x60);
    outportb(dport_lpt1, tempa);
    tempa=tempa & L_L;
    outportb(dport_lpt1, tempa);
}

void end_seq(void)
{
    unsigned char tempb;
    tempb=inportb(dport_lpt1) & 0x9f;
    tempb=tempb | (0x60 & H_L);

    /* generate stop sequence */
    outportb(dport_lpt1, tempb);
    tempb=tempb | (0x60 & H_H);
    outportb(dport_lpt1, tempb);
}

int chk_lpt(void)
{
    /*Get LPT1 port addresses */
    dport_lpt1 = peek(0x40,0x08);
    if(dport_lpt1 == 0) return FALSE;
    return TRUE;
}

int chk_adc(void)
{
    unsigned char temp1;
    outportb(dport_lpt1+2, 0x01);
    temp1=inportb(dport_lpt1);
    temp1=temp1 & 0x7e;
    outportb(dport_lpt1, temp1);
    temp1=inportb(dport_lpt1+2);
```

Listing 8.3 (continued)

```
    temp1=temp1 & 0xfe;
    outportb(dport_lpt1+2, temp1);
    delay(10);
    temp1=inportb(dport_lpt1+1);
    temp1=temp1 & 0x80;
    if(temp1 ) return FALSE;
    temp1=inportb(dport_lpt1+2);
    temp1=temp1 | 0x01;
    outportb(dport_lpt1+2, temp1);
    delay(10);
    temp1=inportb(dport_lpt1+1);
    temp1=temp1 & 0x80;
    if(!temp1) return FALSE;
    adc_con_byte[0]=0x8f;
    adc_con_byte[1]=0xcf;
    adc_con_byte[2]=0x9f;
    adc_con_byte[3]=0xdf;
    adc_con_byte[4]=0xaf;
    adc_con_byte[5]=0xef;
    adc_con_byte[6]=0xbf;
    adc_con_byte[7]=0xff;
    return TRUE;
}

/*start ADC conversion and get the result*/
int read_adc(int channel_number)
{
    int  adc_val, temp_val;
    unsigned char chan_num=0, temp1, temp2, temp3, data[20];
    long loop;
    outportb(dport_lpt1+2, 0x01);
    temp1=adc_con_byte[channel_number];
    for(temp2=0; temp2<8; temp2++)
    {
        temp3= (temp1 << temp2) & 0x80;
        temp3=temp3 & 0x81;
        dport_value=inportb(dport_lpt1);
        dport_value=dport_value & 0x7e;
        dport_value=dport_value | temp3;
        outportb(dport_lpt1, dport_value);
        dport_value=dport_value | 1;
        outportb(dport_lpt1, dport_value);
        outportb(dport_lpt1, dport_value);

        /* this is to make the clk 50% duty cycle*/
        /* Duty cycle as measured with a 66 MHz 486 is 48% */
```

Listing 8.3 *(continued)*

```
        dport_value=dport_value & 0xfe;
        outportb(dport_lpt1, dport_value);
    }
    dport_value=dport_value & 0x7f;
    outportb(dport_lpt1, dport_value);
    for(temp2=0; temp2<16; temp2++)
    {
        dport_value = dport_value & 0x7e;
        dport_value=dport_value | 0x01;
        outportb(dport_lpt1, dport_value);
        data[temp2]=( inportb(dport_lpt1+1) & 0x80);
        dport_value=dport_value & 0xfe;
        outportb(dport_lpt1, dport_value);
        outportb(dport_lpt1, dport_value);
    }
    adc_val=0;
    for(temp2=0; temp2<16; temp2++)
    {
        temp_val=( (unsigned int) data[temp2] & 0x00ff) << 8;
        adc_val= adc_val | ( (temp_val ^ 0x8000) >> temp2);
    }
    adc_val=adc_val>> 3;
    return adc_val;
}

void set_up(void)
{
    unsigned char temp, sda_val=0x50, tempx;
    /*sda_val is set for AD1=AD0=0*/
    unsigned int lp_count;

    /* Read DATA port */
    temp=inportb(dport_lpt1);

    /* Send Slave address byte */
    for (lp_count=0; lp_count<8; lp_count++)
    {
        /* Send SDA */
        temp=temp & L_L;
        tempx=sda_val & 0x80;
        tempx = (tempx >> 2) & 0x20;
        temp = tempx | temp;
        outportb(dport_lpt1, temp);

        /* setup SCL */
        temp = temp | 0x40 ;
```

Listing 8.3 (continued)

```
        outportb(dport_lpt1, temp);

        /* reset SCL */
        temp = temp &  L_H;
        outportb(dport_lpt1, temp);

        /* get new value for SDA */
        sda_val = sda_val <<1;
    }

    /* Send ack */
    temp = temp & 0x9f;
    outportb(dport_lpt1, temp);
    temp = temp | 0x40;
    outportb(dport_lpt1, temp);
    temp = temp & 0x9f;
     outportb(dport_lpt1, temp);
}

void out_to_dac(unsigned int dac_number, unsigned int dac_value)
{
    unsigned char dac_address, tempy;
    unsigned int counter, value, temp1;

    temp1=inportb(dport_lpt1);
    dac_address = 0x07 & ( (char) dac_number) ;
    value = (char) dac_value;

    /* Send command byte to MAX-521 */
    /* Set DAC address into MAX-521 */
    for(counter=0; counter<8; counter++)
    {
        temp1 = temp1 & L_L;
        tempy=dac_address & 0x80;
        tempy = (tempy >> 2) & 0x20;
        temp1 = tempy | temp1;
        outportb(dport_lpt1, temp1);
        temp1 = temp1 | 0x40;
        outportb(dport_lpt1, temp1);
        temp1 = temp1 & L_H;
        outportb(dport_lpt1, temp1);
        dac_address = dac_address << 1;
    }
    /* Send ack */
    temp1 = temp1 & 0x9f;
    outportb(dport_lpt1, temp1);
```

Listing 8.3 (continued)

```c
    temp1 = temp1 | 0x40;
    outportb(dport_lpt1, temp1);
    temp1 = temp1 & 0x9f;
    outportb(dport_lpt1, temp1);

    /* Send value to the selected DAC */
    for(counter=0; counter<8; counter++)
    {
        temp1 = temp1 & L_L;
        tempy = value & 0x80;
        tempy = (tempy >> 2) & 0x20;
        temp1 = tempy | temp1;
        outportb(dport_lpt1, temp1);
        temp1 = temp1 | 0x40;
        outportb(dport_lpt1, temp1);
        temp1 = temp1 & L_H;
        outportb(dport_lpt1, temp1);
        value = value << 1;
    }
    /* Send ack */
    temp1 = temp1 & 0x9f;
    outportb(dport_lpt1, temp1);
    temp1 = temp1 | 0x40;
    outportb(dport_lpt1, temp1);
    temp1 = temp1 & 0x9f;
    outportb(dport_lpt1, temp1);
}

void main(void)
{
    clrscr();
    printf("\nMulti-channel Analog I/O for the PC");
    printf("\n8 Channel 12-bit ADC");
    printf("\n8 Channel 8-bit DACs");
    printf("\nDhananjay V. Gadre, 1996.\n\n");
    if( chk_lpt() == FALSE)
        {printf("\nNo Parallel Port. Aborting..."); exit(1);}
    outportb(dport_lpt1, 0xff);

    if( chk_adc() == FALSE)
        {printf("\nNo ADC Connected. Aborting..."); exit(1);}

    /*Convert voltage on  ADC channel*/
    printf("\nADC Channel 2 Value = %d mV", read_adc(2) );
    printf("\n\nProgramming the 8 DACs..");
    /*Program the DACs*/
```

Listing 8.3 (continued)

```
/* Generate start sequence for MAX-521*/
st_seq();

/*setup address*/
set_up();

/*output to the DACs*/
out_to_dac(0, 0);
out_to_dac(1, 0x20);
out_to_dac(2, 0x40);
out_to_dac(3, 0x60);
out_to_dac(4, 0x80);
out_to_dac(5, 0xa0);
out_to_dac(6, 0xc0);
out_to_dac(7, 0xe0);

/* generate stop sequence */
end_seq();
}
```

## Chapter 9

Expanding Port Bits of the Parallel Port

For many complex applications, the 17 I/O bits that the parallel adapter offers may be insufficient. There are many ways to expand the I/O ports — the choice of a solution may depend on whether you're using the parallel adapter in the standard form or the enhanced form. Although the amount of extra hardware required to add I/O ports to the parallel adapter remains more or less the same for the standard adapter and the enhanced adapter, the software overheads are slightly larger for the standard adapter expansion.

The following sections present three case studies that demonstrate the various ways to expand the I/O on the parallel adapter. In the preceding chapter, I touched on the topic of I/O expansion, specifically in case of the 8-bit interface package. This chapter will explicitly present and explain the expansion mechanism.

The first example offers four digital output ports. The second example offers two digital output and two digital input ports. Each port has eight bits. The first and second examples use the parallel adapter in the standard form. The third scheme shows how to add up to eight digital output ports to an enhanced parallel adapter. This circuit can be easily modified to add digital input ports. The third example interfaces an 8255-PPIO chip to an enhanced parallel adapter. The 8255 chip has three digital I/O ports. This scheme offers true software programmability in the sense that all three ports can be configured in any mode (input or output) without introducing extra hardware.

Expansion on the Standard Parallel Adapter

In this section, I'll examine some schemes for using the parallel adapter to add more digital input and output ports. One way to expand the standard parallel adapter is to think of the DATA port signals as some sort of data bus that can be connected to the inputs of latch ICs. The problem is then to generate strobe pulses for these latch ICs. In this configuration, the CONTROL port signals can be used to generate the trigger (strobe) pulses for the latch ICs. With four CONTROL port signals, four latches can be triggered directly to capture the data provided by the signals of the DATA port. This is shown in the block diagram in Figure 9.1.

Figure 9.1 **A simple scheme to add four digital output ports to the standard parallel adapter.**

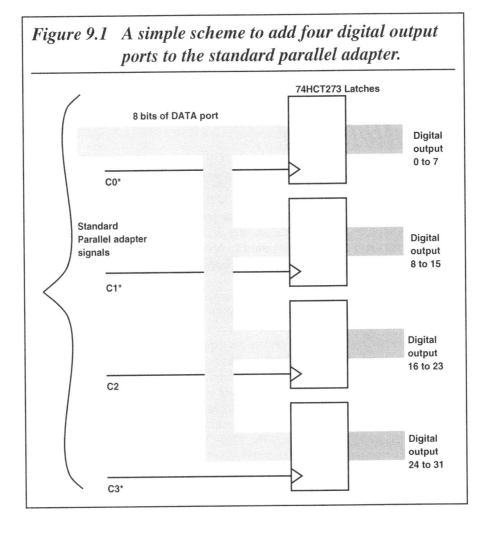

In Figure 9.1, four latch ICs like 74HCT273 (octal latch) are used. The inputs to these ICs are connected to the DATA port signal pins. These latch ICs have a clock input pin. A rising edge on this pin latches the input at the input pins into the latch, and this latched value is available on the output pin until another rising edge is generated on the clock input pin. Each of the four CONTROL port outputs is connected to the clock input pin of the latch IC. When you want to transfer data into one of the latches, the required data is output on the DATA port and the respective CONTROL port pin is pulsed once to latch the data into that latch. Listing 9.1 shows the code for this scheme.

If you need to add more than four ports, the solution is to connect the CONTROL port outputs to a decoder. With four CONTROL port pins, you could use a 4-to-16 decoder and add up to 15 digital output ports.

Listing 9.1   *Provide extra digital output ports on the standard parallel port.*

```
/*expandop.c*/
/*program to provide extra digital output ports (max 4) on the
 standard parallel port.*/

#include <stdio.h>
#include <dos.h>
#include <conio.h>
#include <process.h>

#define BUF_DEPTH 100 /* 100 data points to send to each port*/

/*Global variables that store the addresses of the three ports of the
 standard printer adapter*/
unsigned int dport, cport, sport;

/*routines to generate pulse on each of the 4 control port pins*/
void pulse_c0(void);
void pulse_c1(void);
void pulse_c2(void);
void pulse_c3(void);

void pulse_c0(void) /*generates a low to high pulse on C0 pin*/
{
    unsigned char temp;
    temp=inportb(cport);
    temp=temp | 0x01;
    outportb(cport, temp);
    delay(1);
    temp=temp & 0xfe;
```

The following example shows how to use a decoder to add two digital output ports and two digital input ports. Figure 9.2 shows the block diagram of this expansion scheme. The standard adapter is only equipped to read five bits at a time. A previous

Listing 9.1 (continued)

```
    outportb(cport, temp);
    delay(1);
}

void pulse_c1(void) /*generates a low to high pulse on C1 pin*/
{
    unsigned char temp;
    temp=inportb(cport);
    temp=temp | 0x02;
    outportb(cport, temp);
    delay(1);
    temp=temp & 0xfd;
    outportb(cport, temp);
    delay(1);
}

void pulse_c2(void) /*generates a low to high pulse on C2 pin*/
{
    unsigned char temp;
    temp=inportb(cport);
    temp=temp & 0xfb;
    outportb(cport, temp);
    delay(1);
    temp=temp | 0x04;
    outportb(cport, temp);
    delay(1);
}

void pulse_c3(void) /*generates a low to high pulse on C3 pin*/
{
    unsigned char temp;
    temp=inportb(cport);
    temp=temp | 0x08;
    outportb(cport, temp);
    delay(1);
    temp=temp & 0xf7;
    outportb(cport, temp);
    delay(1);
}
```

Listing 9.1 (continued)

```c
main()
{
    /*the following array stores data to be transfered to the external
      ports, port_0, port_1, port_2 und port_3*/
    unsigned char port_0_buf[100], port_1_buf[100],port_2_buf[100],
            port_3_buf[100], temp;
    unsigned int count;
    /*Get LPT1 port addresses */
    dport = peek(0x40,0x08);
    if(dport ==0)
        {
        printf("\n\n\nLPT! not available... aborting\n\n\n");
        exit(1);
        }
    printf("\nLPT1 address = %X", dport);
    cport = dport +2;       /* control port address */
    sport = dport + 1;      /* status port address */

    /*this statement puts all the CONTROL port signals to logic 1*/
    outportb(cport, 0x04);

    /*setup a loop to transfer the required data points*/
    for(count=0; count<BUF_DEPTH; count++)
            {
            /*transfer data to port 0*/
            /*first output data to the DATA port*/
            outportb(dport, port_0_buf[count]);
            /* now generate a pulse on CONTROL port pin C0 */
            pulse_c0();
            /*transfer data to port 1*/
            /*first output data to the DATA port*/
            outportb(dport, port_1_buf[count]);
            /* now generate a pulse on CONTROL port pin C1 */
            pulse_c1();
            /*transfer data to port 2*/
            /*first output data to the DATA port*/
            outportb(dport, port_2_buf[count]);
            /* now generate a pulse on CONTROL port pin C2 */
            pulse_c2();
            /*transfer data to port 3*/
            /*first output data to the DATA port*/
            outportb(dport, port_3_buf[count]);
            /* now generate a pulse on CONTROL port pin C3 */
            pulse_c3();
            }
}
```

chapter discussed how to connect an 8-bit ADC to the parallel adapter using tristate buffers to interface the 8-bit ADC data with the four STATUS port bits.

Figure 9.2 **Another scheme to add three digital output ports and two digital input ports to the standard parallel adapter.**

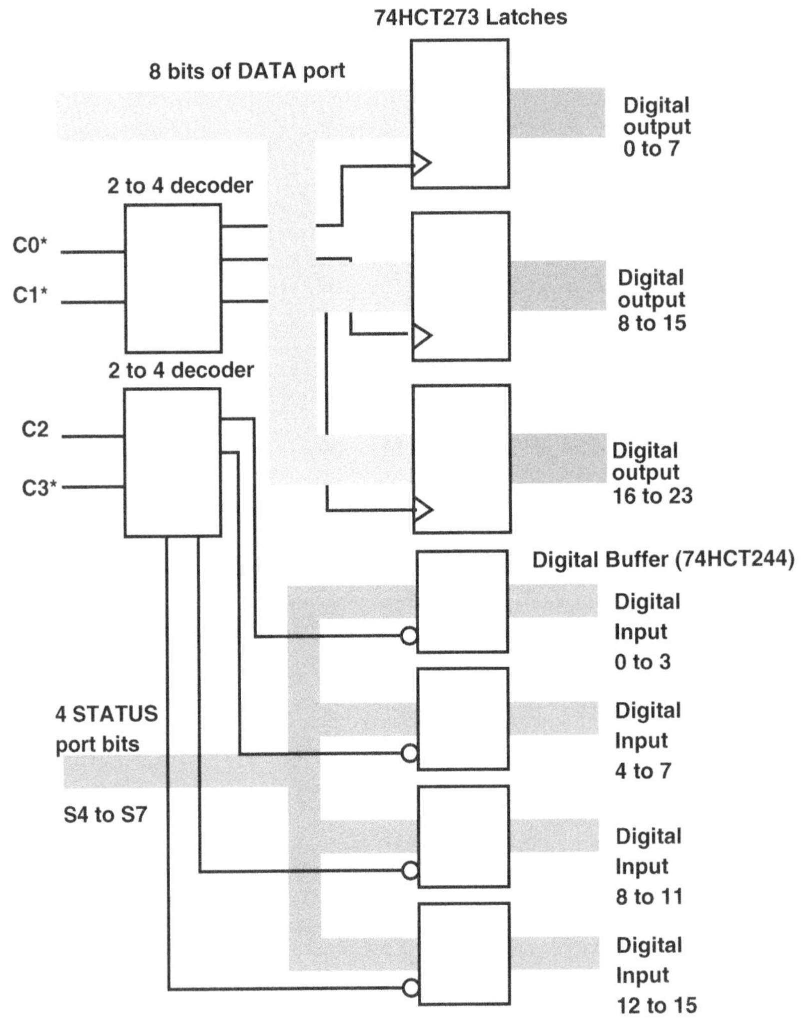

The present example uses a pair of 2-to-4 decoder ICs (actually it is better to use a 3-to-8 decoder in practice, but I will use the 2-to-4 decoder for the purpose of this illustration). The first decoder is driven by the CONTROL port pins C0* and C1*, and the other decoder is driven by C2 and C3*. The first decoder outputs generate latching pulses for three output latch ICs (like the ones used in the last example).

The second decoder (driven by the C2 and C3* pins of the CONTROL port) gen erates the enable signal for the four sections (two each in a 74HCT244 IC) of the tristate buffer ICs. Each tristate buffer section has four input pins. The outputs of the buffers are shorted to create four signals in all, and these are connected to the STA-TUS port pins. The four outputs of the decoder will enable any one of the four buffer sections, and if the program reads the STATUS port, the inputs of that buffer section will be read into the STATUS port.

Note that with four outputs of the decoder, I have connected four tristate buffer sections, but with four other decoder outputs, I connected only three latch ICs. I can-not connect four latches in such a configuration because I must allow one unused out-put, which will serve as the default output signal. If I connect a latch to this output too, data will be written into this latch. The same principle applies if I were to connect a 4-to-16 decoder. I would be able to use only 15 latches.

Expansion Using EPP

The main disadvantage in using the standard parallel port for digital input and output is that the software overheads result in reduced data transfer speeds. You learned in Chapter 5 that the EPP is well suited for data transfer at respectable data transfer rates. Typical data input or output on a 486 machine with a 66MHz clock rate can be close to 900Kbps. In this section, you will see how to use the EPP for digital I/O expansion.

The best way to understand the EPP interface is to think of it as a general-purpose, multiplexed, data and address bus. Normally, data is transferred out as data, but the same lines can also be used to transmit the address of the destination device. For expanding digital I/O, the trick is to trap the address of the destination device and decode it before generating the data strobe.

In the block diagram in Figure 9.3, a few latches are shown with their inputs con-nected to the output of the DATA port signals. The DATA port signals carry data as well as destination port address information. The PC program can choose to write an address to the adapter or it can choose to write the data. You can also write an address byte followed by a data byte, or you can write an address byte followed by reading a data byte. In addition to the data output latches, Figure 9.3 shows an address latch. The address latch has the DATA port connected to its inputs and the clock input con-nected to the address strobe signal (actually, the address strobe signal is generated by ORing the nADDRSTB and nWRITE signal). This arrangement traps any address that is sent on the EPP. The output of the address latch is connected to the input of a suitable decoder. The other input to the decoder is the data strobe signal (which in practice is generated by ORing the nDATASTB and the nWRITE signals). The data latches have their clock inputs connected to each output of the decoder. To write data into any data latch, the program first sets the address for that latch into the address latch and then sends the data on the DATA port pins.

Similar schemes can read data from many digital input ports. Figure 9.4 shows the circuit diagram for a four-port digital output expansion using the EPP. The software for this scheme can be easily modified using code segments from the next section or those presented in Chapter 5.

An 8255-PIO Interface for the EPP

Now that you know how to expand the digital I/O bits of the parallel port using discrete components, it is worthwhile to see if you can do the same thing using programmable I/O devices like the venerable and ubiquitous 8255-PPIO chip.

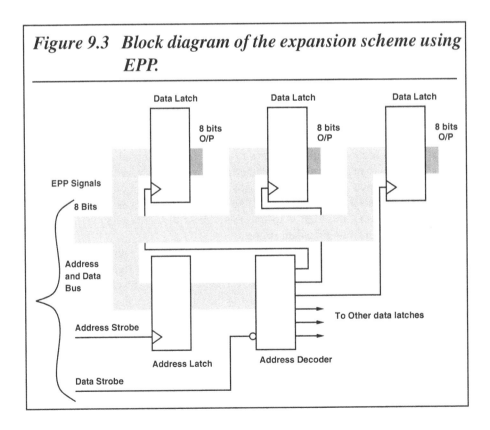

Figure 9.3 Block diagram of the expansion scheme using EPP.

Figure 9.4   Circuit to add extra digital output ports using the EPP.

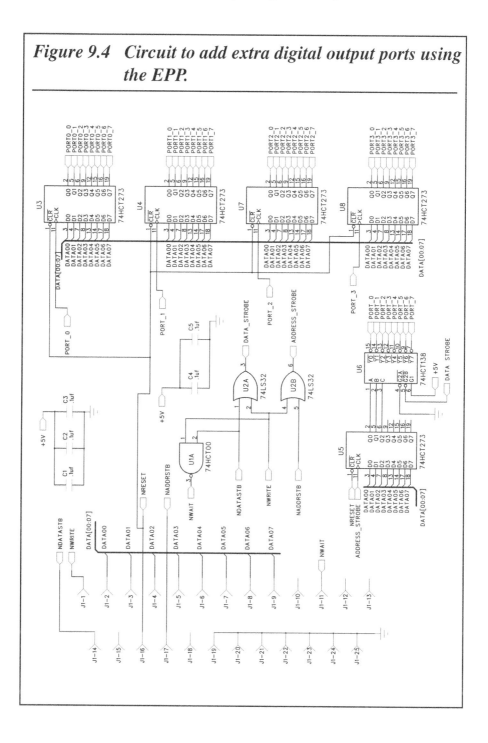

The advantage of using the 8255 chip is that it can be programmed to act as input or output without any changes in the hardware. Thus, a single 8255 configuration can be used in a variety of applications with various I/O requirements.

A look at the details of the 8255 chip will give you a better idea its capabilities. In this example I will use the parallel adapter in EPP mode using the BIOS routines discussed in Chapter 5. However, it would be a simple matter to replace the BIOS routines with code that accesses the parallel adapter chip directly or that uses the parallel adapter in the standard form.

The 8255-PPIO

The 8255 programmable peripheral interface is a general-purpose, programmable I/O device originally designed for Intel microprocessors. It is packaged in a 40-pin DIP package and is fully TTL compatible. It has 24 I/O pins that can be individually programmed in two groups of 12 bits each. To the user, these 24 lines are like three ports of eight bits each. These ports are called port A, port B, and port C. These 24 bits can be used in three major modes of operation.

The three modes are:

- Mode 0: Each group of 12 I/O bits can be programmed in sets of four bits, each to be used as input or output.

- Mode 1: Each group can be programmed to have eight lines of input or output and three of the four remaining lines of that group provide strobe, handshake, and interrupt functions. The other line can be used as a general I/O line.

Table 9.1   Basic operations of the 8255-PPIO chip.

Operation	*A1*	*A0*	*RD**	*WR**	*CS**
Port A read	0	0	0	1	0
Port B read	0	1	0	1	0
Port C read	1	0	0	1	0
Port A write	0	0	1	0	0
Port B write	0	1	1	0	0
Port C write	1	0	1	0	0
Control port write	1	1	1	0	0
Data bus tristate	X	X	X	X	1
Illegal condition	1	1	0	1	0
Data bus tristate	X	X	1	1	0

- Mode 2: This is a bidirectional mode of operation with eight bits as the bidirectional bus. It uses five lines from the second group for handshaking. The second group's eight lines can be used in mode 0 or mode 1 and three lines of the first group can be used as general I/O lines (or handshake lines for lines of the second group). All three of the ports can be programmed in mode 0. Only port A and B can be used for mode 1. If ports A and B are used in mode 1, port C provides the supporting function to ports A and B. Mode 2 is only applicable to port A.

The 8255 chip has an internal data bus buffer that allows the chip to be connected to the system data bus. The data bus is used to transmit and receive data to and from the CPU and to exchange control and status information with the CPU. The data bus buffer transfers this information to the control/status unit (port A, port B, or port C, depending on the state of the A0, A1, WR*, and RD* lines, as shown in Table 9.1).

The block diagram in Figure 9.5 shows the internal scheme of the 8255 chip. The chip is accessed only when the chip select pin, CS*, is held low. Additionally, data is

Figure 9.5 Block diagram of the 8255-PPIO chip.

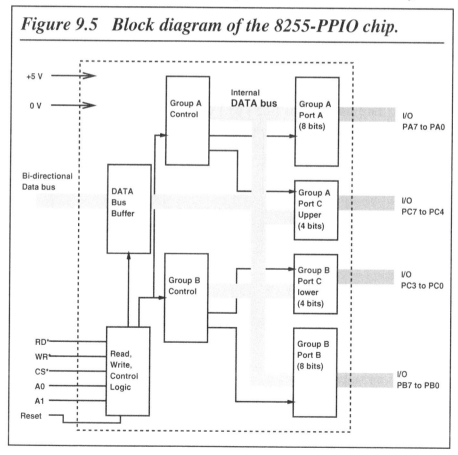

read by the CPU from the 8255 chip when the read pin RD* is also taken low. This pin is used when the CPU wants to read status data or input data from the 8255 chip. When the CPU wants to write control data or output data to the 8255, the write pin WR* is taken low (instead of the RD* pin).

To access the many ports inside the 8255 chip, the chip has two address lines, A0 and A1. Table 9.1 shows the operation of the 8255 chip in response to the signal on its pins.

The three ports of the 8255 chip can be programmed as input or output using the many features described earlier in this section. The individual bits of port C can be set and reset under program control. This very useful feature is required when digital signals are used to control devices such as LEDs, lamps, or relays. With this feature, any bit on port C can be set or reset without bothering with retaining the state of the other bits. this mode of operation is called bit set/reset mode, and it only applies to port C bits.

Figure 9.6 Control word format to program the 8255 chip in the various modes.

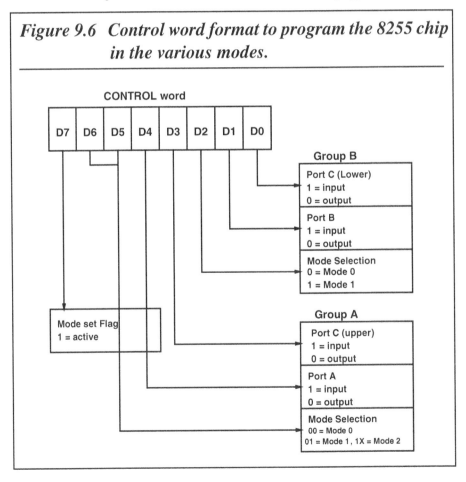

To use the ports of the 8255, the user program must write an appropriate control word into the control register of the 8255. Thereafter, the individual ports can be accessed as desired. Figure 9.6 shows the format of the control byte.

Listing 9.2 shows how to program the 8255 in mode 0 with ports A and B as inputs and port C as output. The code is written with the assumption that the 8255 is interfaced in the I/O port map of the PC at the base address 0x300h, as listed in the code. The base address of 0x300h corresponds to port A, 0x301h to Port B, and 0x302h to port C. The last address would be 0x303h, which would be the address of the control register.

The 8255 supports many modes of operation. This is because of the large number of combinations of the three ports that the user can independently program. It is useful to refer to the 8255 data sheets for complete details on the various modes of operation. The bit set/reset mode is another interesting feature of the 8255. Bit set/reset mode is applicable only to port C of the 8255. Figure 9.7 shows the various combina-

---

Listing 9.2   I/O mode with port A and B as inputs and C as output.

```
/*8255 in I/O mode with Port A and B as Inputs and C as output*/
/*8255_io.c*/
#include<dos.h>
#include<stdio.h>

#define BASE_ADDRESS 0x300

main()
{
   unsigned char control_byte, temp;

   control_byte=0x92;                     /*10010010*/
   /*This control byte programs the 8255 in the I/O mode with all the
   ports in mode 0. Port A and B as inputs and C as output*/

   outportb(BASE_ADDRESS+3, control_byte);
   /*this loads the control register of the 8255 with number 0x92*/

   /*Now any of the 8255 ports can be used.*/
   /*to read the port A and B just do the following*/
   temp= inportb(BASE_ADDRESS);         /*this reads port A*/
   temp = inportb(BASE_ADDRESS + 1);    /*this reads port B*/
   outportb(BASE_ADDRESS+2, temp);
   /*this write the contents of temp into port C*/
}
```

tions of control words that set or reset each and every bit of port C. Listing 9.3 is a piece of code that generates pulses on all the bits of port C one after the other.

The code is written with the assumption that the 8255 is interfaced in the I/O port map of the PC at the base address 0x300h, as listed in the code. The base address 0x300h would correspond to port A. The last address used would be 0x303h, which is the address of the control register.

Parallel Adapter Interface to 8255

The last step is to connect the 8255-PPIO chip to the parallel adapter. This section describes the circuit diagram to connect the 8255 to an EPP. and Listing 9.4 implements the interface using BIOS calls.

Figure 9.8 shows the circuit schematic to interface an 8255-PPIO to an EPP. The circuit shows the EPP signals on the various pins of the 25-pin D male connector. The

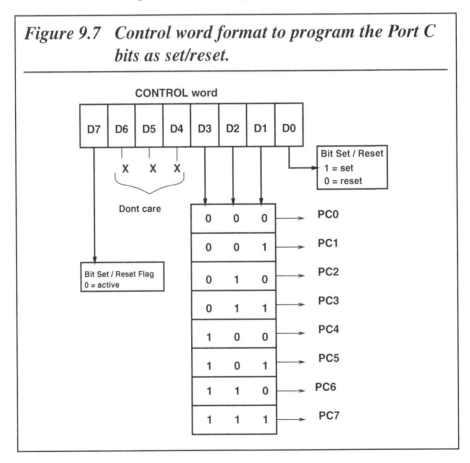

Figure 9.7   Control word format to program the Port C bits as set/reset.

DATA port signals are referred to as DATA00–DATA07 signals, which are connected to the 8255 data input/output pins labeled D0–D7, as well as to the input of the 74HCT273 address latch. The CLR input of this latch is connected to the nRESET sig-

---

Listing 9.3   Generate walking pulses on all the port C bits.

```
/*Program to generate walking pulses on all the port C bits*/
/*8255_exp.c*/
#include<dos.h>
#include<stdio.h>


#define BASE_ADDRESS 0x300


main()
{
    unsigned char control_byte[16], temp;
    /*This buffer stores the control byte set a bit and then to reset it*/
    /*For 8 bits there are 16 bytes*/

    control_byte[0]=0x01; /*set bit 0*/
    control_byte[1]=0x00; /*reset bit 0*/
    control_byte[0]=0x03; /*set bit 1*/
    control_byte[1]=0x02; /*reset bit 1*/
    control_byte[0]=0x05; /*set bit 2*/
    control_byte[1]=0x04; /*reset bit 2*/
    control_byte[0]=0x07; /*set bit 3*/
    control_byte[1]=0x06; /*reset bit 3*/
    control_byte[0]=0x09; /*set bit 4*/
    control_byte[1]=0x08; /*reset bit 4*/
    control_byte[0]=0x0b; /*set bit 5*/
    control_byte[1]=0x0a; /*reset bit 5*/
    control_byte[0]=0x0d; /*set bit 6*/
    control_byte[1]=0x0c; /*reset bit 6*/
    control_byte[0]=0x0f; /*set bit 7*/
    control_byte[1]=0x0e; /*reset bit 7*/

    while(!kbhit()) /*continue till the keyboard is pressed*/
        {
        for(temp=0; temp<16; temp++)
        outportb(BASE_ADDRESS+3, control_byte[temp]);
        /*it must be noted that to set or reset the port C bits, using the
         bit set/reset mode, the information must be written into the
         control register and NOT the port C*/
        }
}
```

Figure 9.8 Schematic for an 8255 interface to an EPP.

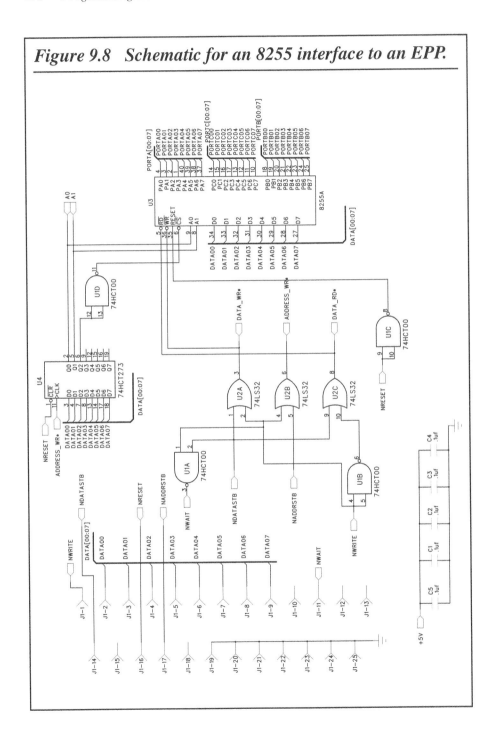

nal of the EPP. The CLK input for the latch is labeled address_wr* and is generated by ORing the nADDRSTB and the nWRITE signal using the 74HCT32 OR gate. The address latch outputs labeled A0 and A1 provide the address A0 and A1 input to the 8255 chip. The CS* signal for the 8255 chip is generated by inverting the Q2 output of the address latch. This is ensure that, at reset, the latch output would be 0, and with the inverter (using the 74HCT00 NAND gate), the CS* will be at logic 1 and will keep the 8255 chip deselected. This will avoid possible contention on the DATA port signals.

The RD* and WR* input signals for the 8255 chip are generated by ORing the nDATASTB with the complement of the nWRITE signal and nDATASTB and nWRITE signals, respectively. The reset input for the 8255 is generated by inverting the nRESET signal output of the EPP.

To transfer data to the 8255, the address for the 8255 port is first latched into the address latch. In the next cycle, the PC writes (or reads) the data to (or from) the designated port. Figure 9.9 shows the logic analyzer trace of the various EPP signals during a composite address write + data write EPP BIOS call. This was captured from a 486 PC running at 66 MHz clock speed.

Listing 9.4 shows the various routines to interface the 8255-PPIO to an EPP using BIOS calls. It would not be difficult to generate the similar code for a standard parallel port or for an enhanced port accessed directly through the controller chip. The connection between the parallel adapter and the 8255 need not change, even if you decide to use different driver software to use the 8255.

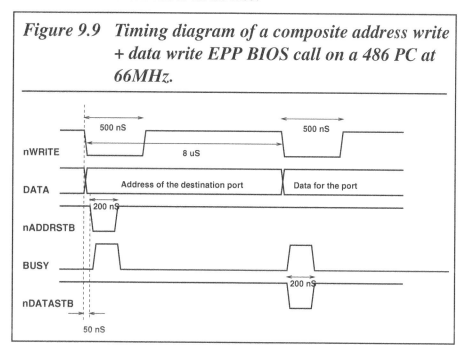

Figure 9.9   Timing diagram of a composite address write + data write EPP BIOS call on a 486 PC at 66MHz.

Listing 9.4 Implement 8255 interface.

```
/* 8255_epp.c */

/*
   Set of Routines for Parallel Port expander using the Enhanced Parallel
   Parallel Port protocol for 8255-PPIO. The routines invoke EPP BIOS calls.

   Dhananjay V. Gadre
*/

#include<dos.h>
#include<time.h>
#include<stdio.h>
#include<process.h>
#include<conio.h>

#define FALSE 0
#define TRUE 1

/* Define the Port addresses of the 8255-PPIO */

#define PORT_A      0x04
#define PORT_B      0x0c
#define PORT_C      0x14
#define CON_PORT    0x1c

void far (*pointr)();
int epp_config(void);
int epp_write_block(unsigned char *source_ptr, int count);
int epp_writea_writed(unsigned char port_address, unsigned char tx_value);
int epp_writea(unsigned char port_address);
int epp_write_byte(unsigned char tx_value);

int epp_write_byte(unsigned char tx_value)
{
    unsigned char temp_ah;

    _AH=7;
    _DL=0;
    _AL=tx_value;
    pointr();
    temp_ah=_AH;
    if(temp_ah != 0) {return FALSE;}
    return TRUE;
}
```

Listing 9.4 (continued)

```c
int epp_writea(unsigned char port_address)
{
    unsigned char temp_ah;
    _AH=5;
    _DL=0;
    _AL=port_address;
    pointr();
    temp_ah=_AH;
    if(temp_ah != 0) {return FALSE;}
    return TRUE;
}

int epp_writea_writed(unsigned char port_address, unsigned char tx_value)
{
    unsigned char temp_ah;
    _AH=0x0c;
    _DL=0;
    _AL=port_address;
    _DH= tx_value;
    pointr();
    temp_ah=_AH;
    if(temp_ah != 0) {return FALSE;}
    return TRUE;
}

int epp_config(void)
{
    unsigned char temp_ah, temp_al, temp_cl, temp_ch;
    _AX=0x0200;
    _DX=0;
    _CH='E';
    _BL='P';
    _BH='P';
    geninterrupt(0x17);
    temp_ah=_AH;
    temp_al=_AL;
    temp_ch=_CH;
    temp_cl=_CL;

    if(temp_ah != 0) return FALSE;
    if(temp_al != 0x45) return FALSE;
    if(temp_ch != 0x50) return FALSE;
    if(temp_cl != 0x50) return FALSE;
```

Listing 9.4 (continued)

```c
    pointr = MK_FP(_DX , _BX);
    _AH=1;
    _DL=0;
    _AL=0x04;
    pointr();
    temp_ah=_AH;
    if(temp_ah != 0) return FALSE;
    return TRUE;
}

int epp_write_block(unsigned char *source_ptr, int count)
{
    unsigned char temp_ah;
    _SI=FP_OFF(source_ptr);
    _ES=FP_SEG(source_ptr);
    _AH=8;
    _DL=0;
    _CX=count;
    pointr();
    temp_ah=_AH;
    if(temp_ah != 0) {printf("\nBlock write timeout error"); return FALSE;}
    return TRUE;
}

main()
{
    int  ret_value, ret_val;
    time_t start, end;
    unsigned char buf_out[10000], rx_in;

    clrscr();
    printf("\nPort Expander using 8255-PPIO and EPP protocol\n");
    ret_value = epp_config();
    if(ret_value == FALSE) { printf("\nNo EPP BIOS"); exit(1);}

    printf("\nEPP BIOS Present");
    printf("\nGenerating a Square wave on Port C bits of the PPIO");
    printf("\nSetting address and transmitting a data byte");

    if( epp_writea(CON_PORT) == FALSE )
        {printf("\nSet address failed.."); exit(1);}

    /* writing 0x80 to Control port puts all the three port of the
       8255 in Output mode */
    if(epp_write_byte(0x80) == FALSE)
        {printf("\nWrite command byte failed.."); exit(1);}
```

Listing 9.4 (continued)

```
    while(!kbhit())
    {

        if(epp_writea_writed(PORT_C, 0) == FALSE)
            {printf("\nWrite data byte failed.."); exit(1);}

        if(epp_writea_writed(PORT_C, 0xff) == FALSE)
            {printf("\nWrite data byte failed.."); exit(1);}

}

/* Another way to access the 8255 data ports. Set the destination
   address once and then keeping writing data to the destination port */

/*Set Port B Address*/
if( epp_writea(PORT_B) == FALSE )
    {printf("\nSet address failed.."); exit(1);}

while(!kbhit())
{

    /*write 0x00 to the Port B of the 8255*/
    if(epp_write_byte(0) == FALSE)
        {printf("\nWrite data byte failed.."); exit(1);}

    /*Write 0xff to the Port B of the 8255*/
    if(epp_write_byte(0xff) == FALSE)
        {printf("\nWrite data byte failed.."); exit(1); }

}

/* Transferring data to Port B using block transfer mode */
printf("\n\nWriting Data Block to the last address");
if (epp_write_block(buf_out, 10000) == TRUE )
    printf("\nBlock write successful");
else printf("\nBlock write failed");

    return 1;
}
```

Chapter 10

Using the Parallel Port to Host an EPROM Emulator

The PC is an ideal machine for hosting microprocessor development tools. The reason for this is the availability of a wide variety of assemblers, compilers, and simulators — programs that are essential for developing microprocessor code.

Imagine you want to develop a microprocessor-based circuit for remote data acquisition. Figure 10.1 shows the block diagram for this application. The components of the circuit are:

- microcontroller or a microprocessor. Typically, a microcontroller is used because it offers increased hardware features, such as timers/counters, I/O ports, on-chip memory, etc.;

- program and data memory. The program memory could be an EPROM. The data memory is RAM IC of sufficient capacity to store the data;

EPROM stands for Erasable Programmable Read Only Memory. EPROM ICs are memory devices that can be programmed with the required code and can also be erased by exposing the chip to UV radiation through the chip's quartz window. You can program the EPROM chip using a special device programmer called an EPROM programmer. To erase the chip, you need a special UV light source.

- communication link to the host computer to upload the parameters and, subsequently, to download the recorded data to the host;

- digitizer or an ADC that converts the sensor outputs to bits that the microprocessor reads and stores in the data memory;

- power conditioning circuit to convert the raw battery voltage to levels required by the various components of the circuit.

The steps in developing this application include:

1. formulate the requirements of the circuit in terms operational features, required hardware, power consumption, etc.;

2. finalize the hardware design with a few iterations;

3. draw a flowchart of the software showing major routines, etc.;

4. write preliminary code and test the code on a simulator if one is available;

5. fabricate a hardware prototype board;

6. use a cross compiler or cross assembler for the selected microprocessor;

7. burn the executable code in an EPROM and test it on the prototype; in case of errors, correct the code, erase the EPROM, and burn the new code into the EPROM again; and

8. iterate from step six until the system starts working without errors.

For a system as complex as this example, the software could take a few thousand assembler lines and is not a simple affair. In a system as big as this, using this method involves a lot of time for debugging code, erasing and burning new code in the

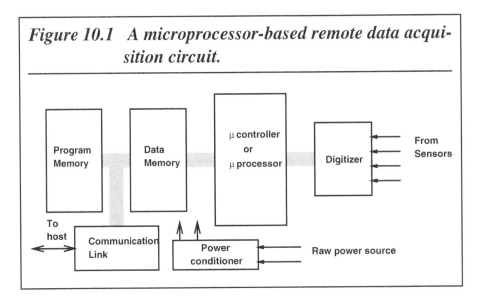

Figure 10.1   A microprocessor-based remote data acquisition circuit.

EPROM, and testing the system. Mistakes are bound to occur, and it is better to assume that mistakes will occur and be prepared to tackle them. Some tools that help circumvent the lengthy iterations of burning and erasing EPROMs are discussed in the next section.

Microprocessor Development Using Emulators

The most convenient (as well as the most expensive) tool for designing microprocessor-based circuits and code is a microprocessor (or microcontroller) emulator. A processor emulator, also called an In-Circuit Emulator (ICE) is, as the name suggests, a piece of hardware that mimics the processor. An ICE is usually supplied by the microprocessor manufacturer (but recently, third party companies have begun to offer emulators — sometimes even before the processor manufacturer can offer one). The connector footprint of the emulator is identical to the actual microprocessor and is supposed to be inserted into the socket on the target board, where eventually the actual processor will sit. This emulator hardware has a connection to the PC (called the umbilical cord by some) that allows test code to be downloaded from the PC into the emulator.

One way to use the ICE as a development tool is to take the bare target circuit board and populate it with the minimum of components — just enough to support an ICE. The ICE is then plugged into this partially populated PCB. Code to test very rudimentary functions is then downloaded into the ICE from the PC host (the PC host is equipped with the necessary editor, assembler, compiler debugger, etc.), and preliminary tests are performed on the circuit. These may just include looking for address strobes on the PCB.

Progressively, more components are soldered onto the PCB, and more test routines are downloaded to perform more tests. Eventually, you reach a point where all the components have been soldered, and physical and logical connections between the various components on the PCB have been tested. Now is the time to integrate the various routines into the complete system software. Here, also, the ICE is of great help. The designer can download version after version of the system software into the ICE until the performance meets the requirements.

The ICE also allows the user to trace each and every instruction and monitor the effect of the instruction execution using any suitable device (a logic analyzer or an oscilloscope). The emulator will execute one instruction and then wait for the user to proceed. When the code is acceptable, the ICE is removed from the socket and the developed code is burned into a suitable ROM/EPROM. After the actual microprocessor is placed in the target socket, the PCB hardware and software are ready.

The other approach to using an ICE is only slightly different from the first approach. The alternative is to put all the components except the processor and the ROM/EPROM in place and test everything with the ICE. Using this approach may discover some problems, such as wrong circuit tracks or bad design, which would be a little difficult to correct with all the components soldered in place.

An ICE is an expensive piece of equipment and is sometimes beyond the reach of many designers. Another disadvantage of an ICE is that it is very specific to the microprocessor you are using. There is no such thing as a universal ICE. For every processor type you have to buy an ICE.

There is a simple solution to this problem, though there are some trade-offs. At the cost of reduced debugging and trace features, this low-cost solution allows emulation and a fraction of the cost of an ICE. The solution is to use an EPROM/ROM emulator instead of a processor emulator.

Figure 10.2 shows the block diagram of SmartRAM, an extremely simple EPROM emulator. SmartRAM can emulate 2Kb EPROMs, which may look tiny by prevailing standards, but the idea behind SmartRAM can be extended to support emulation of larger capacity EPROM and ROM devices.

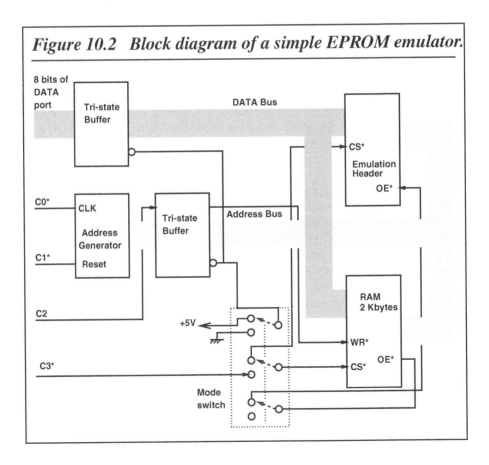

Figure 10.2   Block diagram of a simple EPROM emulator.

SmartRAM with 2Kb EPROM support is more than enough for relatively smaller applications that are programmed using assembly language. For larger applications, 2Kb of memory may be insufficient and you may need to augment the capacity of SmartRAM. Also, for a microprocessor application developed using C or another high-level language, 2Kb of memory may be too small, and you may need to think of a higher capacity EPROM emulator. SmartRAM's simple design is open to modifications. Figure 10.2 shows inputs to a tristate buffer connected to the DATA port signal pins. The outputs of the buffer are connected to the data pins of a 2Kb RAM chip. An address generator block provides the address for the RAM chip. The outputs of the address generator are also passed through the tristate buffer IC. The address and data bus of the RAM is also connected to a DIP socket called the emulation header. The connection between the RAM and the DIP header is by means of suitable wires. A three-pole switch is used. One pole of the switch controls the tristate buffers. In one position of the switch, the emulator is put in the emulation mode. In the other position, the emulator operates in the load mode. In the load mode, the tristate buffers are enabled and the address bits from the address generator block connect to the RAM IC. Also, the DATA port signals connect to the data pins of the RAM. Besides this, two control pins of the RAM, the CS* and the WR* signals, get connected to the CONTROL port bits C2 and C3*.

Also, the CS* and OE* signals from the emulation header are disconnected during the load mode of operation. In this mode, the PC program can write values into any of the locations of the RAM. To do so, it first resets the address generator (which makes the address value equal to 0). Then it generates as many clock pulses for the CLK input of the address generator as the address of the RAM chip to which that byte needs to be written. If the program is required to load the value 0x44h in location 167 (decimal), the PC resets the address generator and generates 167 clock pulses. This puts the address value to 167 (the required value). The program then outputs 0x44h on the DATA port signals and generates the CS* and WR* signals. This writes 0x44h into the address location 167 of the RAM.

In the emulation mode of operation, the switch disables all the tristate buffers and connects the OE* and CS* signals from the emulation header to the RAM chip. Because the emulation header is a DIP socket with the same specifications as the EPROM chip it emulates, the DIP socket can be inserted in the target EPROM socket. The microprocessor on the target board can now access the code stored in the RAM chip (which the processor believes is being read out of an EPROM). When you need to change the code, you just have to turn the target PCB power off, remove the emulation socket from the EPROM socket on the target board, and put the switch on the emulator in load mode. This allows new code to be loaded into the RAM, after which the emulator is ready again for emulation.

Using SmartRAM

To use SmartRAM, you need an external +5V power supply that can supply up to 100mA of current. The actual current consumption of the emulator is much less. Figure 10.3 shows a circuit schematic of the SmartRAM EPROM emulator.

The emulator is connected to the parallel adapter connector. The emulation header is disconnected from the target socket. After connecting the power supply, the mode switch is put in the load mode and the SmartRAM program on the PC is executed. This SmartRAM driver program offers the following options:

- L: Load intelhex format object code into the PC buffer memory;
- F startadd endadd constant: Fills the PC buffer memory from address startadd to endadd with a constant value;
- D xx: Dumps buffer memory contents from address xx onward onto the screen;
- W: Writes the emulator RAM with the contents of the PC buffer memory;
- Q: Quits the program.

Using the single letter commands shown above, the user can load intelhex object code, change individual memory contents, load an entire block of memory with a fixed value, and finally, transfer the PC memory buffer into the emulator RAM. At this point, you are ready to use the emulator in emulation mode. To do this, power up the target PCB and insert the emulator into the target EPROM socket. After that the emulator switch is put in emulation mode and the target microprocessor is reset. The microprocessor now starts reading the emulator for the code.

It is important to note that the emulator socket should never be pulled out of a live target PCB while in emulation mode. To remove the emulator socket, turn off the target power and the emulator power, then plug out the emulator header from the EPROM socket.

Driver Software

In the last section we looked at the possible features of the SmartRAM EPROM emulator. Even though I am not providing any driver software for the emulator, it should be possible to code a simple program to control the emulator as sketched in the last section (Listing 10.1).

Figure 10.3 Circuit schematic of the SmartRAM EPROM emulator.

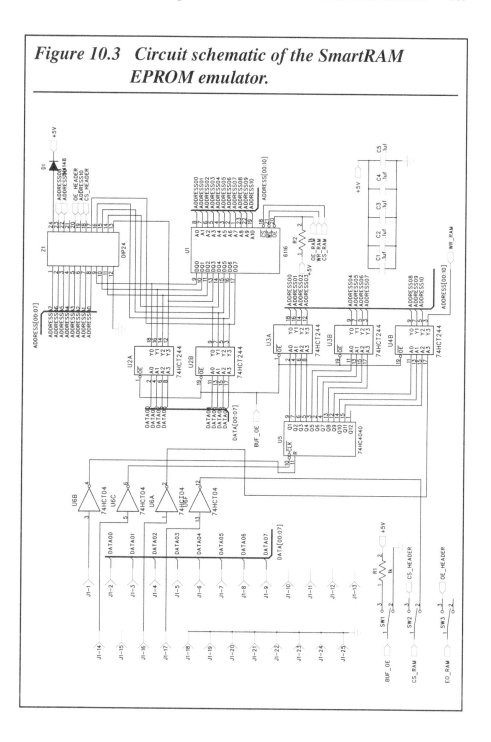

EPROM Emulation Using
Non-Volatile RAM (NVRAM) Modules

As we have seen, EPROM emulation is a fairly established, in-expensive method of code development for embedded processors. Many vendors supply EPROM emulators with multiple device emulation facility; from the measly 2Kb EPROMs to 1Mb devices.

In the last example, we have also seen that EPROM emulators have a host computer link connected to the emulation hardware and a pod that emulates the required EPROM. With a few jumper settings on the emulator besides changing to a suitable pod, the device can be used to emulate a different EPROM.

Listing 10.1   Driver software for the SmartRAM project.

```
/* nvemul.c */

/*
   EPROM Emulator using NVRAM module.
   32Kbyte capacity.
   Dhananjay V. Gadre
   12th July 1997
*/

#include<stdio.h>
#include<conio.h>
#include<dos.h>
#include<process.h>
#include<time.h>
#include<alloc.h>
#include<ctype.h>
#include<stdlib.h>
#include<string.h>

/* port addresses of the parallel adapter */
unsigned int dport, sport, cport;

/* these ports control data, voltage to the ZIF socket, */
/* control the tri-state buffers respectively */
unsigned char port_0, port_1, port_2, port_3, error_byte;

#define MEMORY 32768 /* last address of the target controller memory */
```

However, there is a problem with using such emulators in applications that have cramped PCB real-estate, or in a stacked PCB configuration or under circumstances which do not offer any space to place the host computer or the emulator hardware. These cases may sound bizzare, but they are real. What to do in such circumstances? Well there is a way out of this dilemma. My solution is to use now freely available Non Volatile RAM (NVRAM) module to mimic the required EPROMs. NVRAM modules are made by many manufacturers, including Dallas Semiconductors. The memory module contains the standard RAM chip together with a Lithium battery to retain the memory contents. It is said that the lithium battery can retain the memory contents for upto 10 years from the time the memory is first used in circuit. Thus, it is very easy to use the NVRAM module as a portable data transfer element.

In this section, we look at a new and novel method of EPROM emulation. I describe a scheme to load any required data into a 32Kb NVRAM module using a simple hardware that can be connected to the parallel port. After writing and verifying the contents of the NVRAM, the NVRAM can be removed from the code writer hardware and plugged into the required EPROM socket.

The PC driver software (which is specifically written for EPROM emulation) can read intelhex object files and write the data into the NVRAM module. The data that is written into the module is also verified by the driver software.

Circuit Description

The circuit, RamWriter, uses a 32Kb NVRAM, module, the DS1230 from Dallas Semiconductors as the emulator element. Other similar NVRAM modules can also be used. NVRAM modules are pin compatible with SRAM and EEPROM components but the pin out is a little different from an EPROM device. We will come to this a little later and discuss ways and means to circumvent the problem.

RamWriter connects to the PC host through our dear parallel port. Figure 10.4 shows the circuit diagram for RamWriter. The data to the NVRAM is written by the latch U2 (74HCT573). The WR*, CS* and OE* signals for module are generated by the U1 latch (74HCT273). At power on, all the control signals (OE*, CS* and WR*) are disabled. During data verify, the data is read back by the U9 buffer by the STATUS port of the parallel port.

The address for the module is generated by the U3 and U4 latches (74HCT273). The clock signals for all the latches is generated by the 4 CONTROL port signals C0*, C1*, C2 and C3*. The inputs to all the latches is provided by the output of the DATA port of the parallel port.

LED D2 provides power on indication and D3 provides information when the data is being written to or being read from the module. J2 is a 25 pin D' mail connector that fits directly onto the 25 pin `D' female parallel port connector.

Figure 10.4 Circuit schematic of the NVRAM-based EPROM emulator.

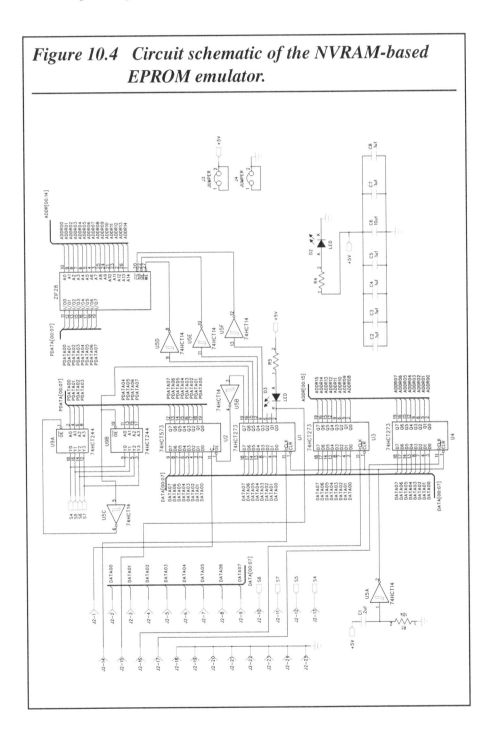

To allow ease of insertion and removal of the module, a 28 pin ZIF socket is used for the NVRAM module. The circuit needs external +5V TTL grade power supply at less than 25 mA current.

Driver Software

The driver software is written in C and compiled under Borland C compiler. The software reads intelhex object files and writes the code into the NVRAM module. It can detect if the module is missing. Any error in data verification is also detected and reported. At the end of the exercise, the circuit power supply is switched off and the NVRAM module is removed from the RamWriter. It can now be used to emulate a 32Kbyte EPROM. Which brings us to the problem of differing pinouts of NVRAM and EPROM. The NVRAM and EPROM share the same pinout except the pins 1 and 27. The following table lists the functions of these two pins for the EPROM and the NVRAM.

Pin #	NVRAM	EPROM
1	A14	Vpp
27	WR*	A14

During the use of the EPROM in a circuit, the Vpp pin is held at +5V. During emulation, the WR* pin of the NVRAM is also held to its inactive state i.e. +5V. So the following modification which can be easily built into any application circuit will allow the use of the NVRAM as an EPROM.

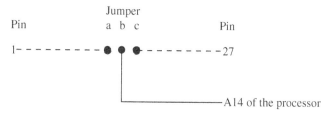

The modification shown above involves the use of a 3 pin jumper switch which needs to be set appropriately. When using the NVRAM, pins a and b of the jumper are shorted. When using EPROM, pins b and c are shorted. Also, pins 1 and 27 are pulled up to +5V with a 10Kohm resistors. These additions can easily be incorporated into the application circuit.

Typical Uses

This EPROM emulator can be used in any application that uses a 8 bit data path to execute the code. Any 8-bit microprocessor and 8 bit controllers with external program memory could benefit from this device. Also, ADSP21xx range of DSP controllers from Analog Devices use EPROM to load the internal program (which is 24-bits) memory from an external byte-wide EPROM at reset. I have used the RamWriter to develop code for a 8031 based application and another ADSP-2105 DSP based application.

Besides using this tool for EPROM emulation, the driver software can be easily modified to read NVRAM contents and store the data into a file in applications where the NVRAM is used as a portable data transfer element as in data loggers etc.

Listing 10.1 (continued)

```
/* the Intelhex file has lines of code. each line begins with a : */
/* next is number of bytes in the line */

/* offset in the hex file where the line length is stored */
#define LL 1

/* offset in the hex file where the destination address is stored */
#define ADDR 3
#define ZEROS 7

/* offset of the beginning of the code */
#define CODE_ST 9

/* local global variables */
unsigned char  ram[MEMORY+10];
unsigned int curr_address;
FILE *fp1;

/* defines for port1 */
#define ENB_DATA 0x80
#define DIS_DATA 0x7F
#define ENB_LOW  0x40
#define ENB_HIGH 0xbf
#define ENB_WR   0x20
#define ENB_OE   0x10
#define ENB_CS   0x09

/* local routines */
int initialze(void);    /* initialzes the external hardware */
```

Listing 10.1 (continued)

```c
/* read the intelhex format file & fill up the internal buffer */
int fill_buffer(void);

/* check if the programmer is connected and if +12V is ON */
int chk_writer(void);
int write_verify_bytes(void);

/* routines to generate pulse on each of the 4 control port pins */
void pulse_c0(void);
void pulse_c1(void);
void pulse_c2(void);
void pulse_c3(void);

void pulse_c0(void)
{
    unsigned char temp;
    temp=inportb(cport);
    temp=temp & 0xfe;
    outportb(cport, temp);
    outportb(cport, temp);
    outportb(cport, temp);
    temp=temp | 0x01;
    outportb(cport, temp);
}

void pulse_c1(void)
{
    unsigned char temp;
    temp=inportb(cport);
    temp=temp | 0x02;
    outportb(cport, temp);
    outportb(cport, temp);
    outportb(cport, temp);
    temp=temp & 0xfd;
    outportb(cport, temp);
}

void pulse_c2(void)
{
    unsigned char temp;
    temp=inportb(cport);
    temp=temp & 0xfb;
    outportb(cport, temp);
    outportb(cport, temp);
    outportb(cport, temp);
    temp=temp | 0x04;
    outportb(cport, temp);
}
```

Listing 10.1 (continued)

```
void pulse_c3(void)
{
    unsigned char temp;
    temp=inportb(cport);
    temp=temp | 0x08;
    outportb(cport, temp);
    outportb(cport, temp);
    outportb(cport, temp);
    temp=temp & 0xf7;
    outportb(cport, temp);
}

char chartoi(char val)
{
    unsigned char temp;
    temp = toupper(val);
    if(temp>0x39) {temp = temp -0x37;}
    else {temp=temp-0x30;}
    return temp;
}

int initialize(void)
{
    dport = peek(0x40, 8);
    sport=dport+1;
    cport=dport+2;
    if(dport ==0) return 0;
    outportb(dport, 0);
    outportb(cport, 0x05);     /* all cport outputs high, except C0 */
    outportb(cport, 0x0a);     /* all cport pins are low, except C0 */
    outportb(cport, 0x05);     /* all cport outputs high, except C0 */
    port_0=0;
    port_1=0;
    port_2=0;
    port_3=0;
    return 1;
}

/* read the intelhex format file & fill up the internal buffer */
int fill_buffer(void)
{
    unsigned char ch, temp4, temp1, temp2, temp3;
    unsigned char chk_sum=0, buf[600], num[10];
    unsigned int line_length, address, line_temp, tempx, count=0;
```

Listing 10.1 *(continued)*

```
count=0;
while(!feof(fp1) )
{
    chk_sum 0,

    /* check if start of line = ':' */
    fgets(buf, 600, fp1);
    tempx=strlen(buf);

    /* printf("\n\nString length=%d\n", tempx); */
    /* printf("\n\n%s", buf); */
    if( buf[0] != ':') {printf("\nError...
    Source file not in Intelhex format. Aborting");
    fclose(fp1);
    return 0;
}

/* convert the next 2 characters to a byte which equals line length */
temp1=buf[LL];
temp2=buf[LL+1];
if( !isxdigit(temp1) ) {
    printf("\nError in source file. Aborting");
    fclose(fp1);
    return 0;
}

if( !isxdigit(temp2) ) {
    printf("\nError in source file. Aborting");
    fclose(fp1);
    return 0;
}

temp4 = chartoi(temp1);
chk_sum=chk_sum + 16*temp4;
line_length=(int)temp4;

temp4=chartoi(temp2);
chk_sum=chk_sum + temp4;
line_length = 16*line_length + (int)temp4;
/* printf("Entries=%d  ", line_length); */

if(line_length ==0) {
    return count;
}
```

Listing 10.1 (continued)

```c
temp1=buf[ADDR];
temp2=buf[ADDR+1];
temp3=buf[ADDR+2];
temp4=buf[ADDR+3];

if( !isxdigit(temp1) ) {
    printf("\nError in source file. Aborting");
    fclose(fp1);
    return 0;
}

if( !isxdigit(temp2) ) {
    printf("\nError in source file. Aborting");
    fclose(fp1);
    return 0;
}

if( !isxdigit(temp3) ) {
    printf("\nError in source file. Aborting");
    fclose(fp1);
    return 0;
}

if( !isxdigit(temp4) ) {
    printf("\nError in source file. Aborting");
    fclose(fp1);
    return 0;
}

ch = chartoi(temp1);
temp1=ch;

ch = chartoi(temp2);
temp2=ch;

chk_sum = chk_sum + 16*temp1 + temp2;

ch = chartoi(temp3);
temp3=ch;

ch = chartoi(temp4);
temp4=ch;

chk_sum = chk_sum + 16*temp3 + temp4;

address = 0x1000 * (int)temp1 + 0x100 *
        (int)temp2 + 0x10*(int)temp3 + (int)temp4;
```

Listing 10.1 (continued)

```c
/* printf("Start Address=%x hex\n", address); */
if(address > MEMORY)
{
    printf("\nError in source file. Bad address. Aborting");
    fclose(fp1);
    return 0;
}

/* check for the next byte. It has to be 00 * */
temp1=buf[ZEROS];
temp2=buf[ZEROS+1];

if( !isxdigit(temp1) ) {
    printf("\nError in source file. Aborting");
    fclose(fp1);
    return 0;
}

if( !isxdigit(temp2) ) {
    printf("\nError in source file. Aborting");
    fclose(fp1);
    return 0;
}

ch = chartoi(temp1);
temp1=ch;
ch=chartoi(temp2);
temp2=ch;
ch = 16*temp1 + temp2;
if(ch != 0)
{
    printf("\nError... Source file not in Intelhex format. Aborting");
    fclose(fp1);
    return 0;
}

/* now read bytes from the file & put it in buffer */
for(line_temp=0; line_temp<line_length; line_temp++)
{
    temp1=buf[2*line_temp+CODE_ST];
    temp2=buf[2*line_temp+CODE_ST+1];

    if( !isxdigit(temp1) ) {
        printf("\nError in source file. Aborting");
        fclose(fp1);
        return 0;
    }
```

Listing 10.1 (continued)

```
        if( !isxdigit(temp2) ) {
            printf("\nError in source file. Aborting");
            fclose(fp1);
            return 0;
        }
    ch = chartoi(temp1);
    temp1=ch;
    ch=chartoi(temp2);
    temp2=ch;
    ch = 16*temp1 + temp2;
    chk_sum=chk_sum + ch;

    if(address > MEMORY)
    {
        printf("\nError in source file. Bad address. Aborting");
        fclose(fp1);
        return 0;
    }

    /* printf("%X ",ch); */
    ram[address]=ch;
    address++;
    count++;
}

/* get the next byte. this is the chksum */

temp1=buf[2*line_length+CODE_ST];
temp2=buf[2*line_length+CODE_ST+1];

if( !isxdigit(temp1) ) {
    printf("\nError in source file. Aborting");
    fclose(fp1);
    return 0;
}

if( !isxdigit(temp2) ) {
    printf("\nError in source file. Aborting");
    fclose(fp1);
    return 0;
}
ch = chartoi(temp1);
temp1=ch;
ch=chartoi(temp2);
temp2=ch;
ch = 16*temp1 + temp2;
chk_sum=chk_sum + ch;
```

Listing 10.1 *(continued)*

```c
    /* printf("csum=%x", chk_sum); */
    if(chk_sum !=0)
    {
        printf("\nChecksum Error... Aborting");
        fclose(fp1);
        return 0;
    }
}
return count;
}


int write_verify_bytes(void)
{
    unsigned char temp, low_temp, high_temp, low_addr, high_addr;
    long program_length;

    for(program_length=0; program_length < MEMORY; program_length++)
    {
        curr_address=program_length;

        /* generate the address for the NVRAM */

        /* this puts the data into the data latch
           but the latch is yet to be enabled */
        port_0=ram[program_length];
        outportb(dport, port_0);
        pulse_c0();

        /* enable the data on the data pins */
        port_1=port_1 | ENB_DATA;
        outportb(dport, port_1);
        pulse_c1();

        /* write data to the NVRAM */
        low_addr=(unsigned char)program_length;
        high_addr=(unsigned char) ((program_length>>8) & 0x00ff);

        port_2=high_addr;
        outportb(dport, port_2);
        pulse_c2();

        port_3=low_addr;
        outportb(dport, port_3);
        pulse_c3();
```

Listing 10.1 (continued)

```
        /* generate chip select */
        port_1=port_1 | ENB_CS;
        outportb(dport, port_1);
        pulse_c1();

        /* generate write strobe */
        port_1=port_1 | ENB_WR;
        outportb(dport, port_1);
        pulse_c1();

        /* disable write strobe */
        port_1=port_1 & ~ENB_WR;
        outportb(dport, port_1);
        pulse_c1();

        /* disable data from the data output pin of the source */
        port_1=port_1 & DIS_DATA;
        outportb(dport, port_1);
        pulse_c1();

        /* now enable read strobe to read back the data */
        port_1=port_1 | ENB_OE;
        outportb(dport, port_1);
        pulse_c1();

        /* read low nibble */
        port_1=port_1 | ENB_LOW;
        outportb(dport, port_1);
        pulse_c1();
        low_temp=inportb(sport);

        /* read high nibble */
        port_1=port_1 & ENB_HIGH;
        outportb(dport, port_1);
        pulse_c1();
        high_temp=inportb(sport);

        /* disable cs strobe */
        port_1=port_1 & ~ENB_CS;
        outportb(dport, port_1);
        pulse_c1();

        temp= (high_temp & 0xf0) | ( (low_temp >>4) & 0x0f);
        temp=temp ^ 0x88;
```

Listing 10.1 (continued)

```
        /* printf("%X ", temp); */
        if(temp != ram[program_length])
        {
            error_byte=temp;
            printf("\nError in program verify at address %X (hex).
                    Aborting...", program_length);
            printf("\nProgram data %X, read back data %X\n",
                    ram[program_length], temp);
            return 0;
        }
    }

    for(program_length=0; program_length < MEMORY; program_length++)
    {
        curr_address=program_length;

        /* generate the address for the NVRAM */

        /* write data to the NVRAM */
        low_addr=(unsigned char)program_length;
        high_addr=(unsigned char) ((program_length>>8) & 0x00ff);

        port_2=high_addr;
        outportb(dport, port_2);
        pulse_c2();

        port_3=low_addr;
        outportb(dport, port_3);
        pulse_c3();

        /* generate chip select */
        port_1=port_1 | ENB_CS;
        outportb(dport, port_1);
        pulse_c1();

        /* now enable read strobe to read back the data */
        port_1=port_1 | ENB_OE;
        outportb(dport, port_1);
        pulse_c1();

        /* read low nibble */
        port_1=port_1 | ENB_LOW;
        outportb(dport, port_1);
        pulse_c1();
        low_temp=inportb(sport);
```

Listing 10.1 (continued)

```
        /* read high nibble */
        port_1=port_1 & ENB_HIGH;
        outportb(dport, port_1);
        pulse_c1();
        high_temp=inportb(sport);

        /* disable cs strobe */
        port_1=port_1 & ~ENB_CS;
        outportb(dport, port_1);
        pulse_c1();

        temp= (high_temp & 0xf0) | ( (low_temp >>4) & 0x0f);
        temp=temp ^ 0x88;
        printf("%X ", temp);
        if(temp != ram[program_length])
        {
            error_byte=temp;
            printf("\nError in program verify at address %X (hex).
                    Aborting...", program_length);
            printf("\nProgram data %X, read back data %X\n",
                    ram[program_length], temp);
            return 0;
        }
    }
    return 1;
}

main(argc, argv)
int argc;
char *argv[];
{
    time_t start, endt;
    unsigned long temp;
    int byte_value, return_val, total_bytes;

    printf("\n\n\n\tEPROM Emulator using NVRAM module Ver: 1.0\n");
    printf("\t----------------------------------------\n");
    printf("\t\t    Dhananjay V. Gadre");
    printf("\n\t\t      July  1997.\n"); /* 12th July  1997 */

    if(argc != 2) {
        printf("\nError... Specify Intelhex source filename. Aborting");
        printf("\nFormat: AtmelP intelhex_sourcefile");
        exit(-1);
    }
```

Listing 10.1 (continued)

```
    if((fp1=fopen(argv[1], "r")) == NULL) {
        printf("\nError...Cannot open source file. Aborting");
        exit(-1);
    }

    return_val=initialize(); /* Initialize the printer adapter port */
    if(return_val == 0) {printf("\nLPT1 not available. Aborting...");
    fclose(fp1);
    exit(-1);
}

printf("\nLPT1 DATA port address = %X (hex)", dport);

printf("\nChecking internal memory buffer...");
for(temp=0; temp < MEMORY; temp++)
{
    ram[temp]=(unsigned char) temp;
}

for(temp=0; temp < MEMORY; temp++)
{
    if( ram[temp] != (unsigned char) temp)
    {
        printf("\nError in internal memory allocation...Aborting.");
        fclose(fp1);
        exit(-1);
    }
}

printf("\nInternal memory buffer OK.");

printf("\nReading Intelhex source file...:");

return_val=fill_buffer();
if(return_val == 0) {
    exit(0);
}
```

Listing 10.1 (continued)

```
printf("\nIntel hex file %s read successful.
        Total bytes read =%d", argv[1], return_val);

fclose(fp1);
printf("\nStoring data in NVRAM module and Verifying...\n");
return_val=write_verify_bytes();
if(return_val == 0) {
    printf("\nFailed to store data in NVRAM at address:
        %X (hex)\n", curr_address);
    printf("Program value: %X\n", ram[curr_address]);
    printf("Verify value: %X\n", error_byte);
    exit(-1);
}

printf("\nData stored in NVRAM  and verified");
printf("\nPower Off the RAMWriter and remove the NVRAM module");
return 1;
}
```

Chapter 11

The Parallel Port as a Host Interface Port

This chapter looks at using the parallel adapter as a channel for communicating between the PC processor and another embedded processor. Many complex control applications must perform intensive mathematical calculations. One approach to tackling this problem is to use the PC as an all-in-one device to provide a user interface, a mathematical calculation engine, and I/O to the external device that needs to be controlled. Another approach is to use the PC only to provide a user interface and to use a separate embedded controller such as a Digital Signal Processor (DSP) to perform the math and I/O. The latter solution is especially attractive because it does not put any demands upon the PC resources (as long as the PC can provide a suitable graphical user interface). Motion control systems typically use this approach. If you are going to interface the PC with a DSP, it is important to provide a suitable communication channel between the PC and the embedded controller.

If the required data transfer rate between the PC and the controller is higher than the rate provided by the serial RS-232 port, the parallel adapter is a good candidate for a communications interface.

This chapter examines the problem of communicating between the PC processor and an external controller through the parallel adapter. Two examples illustrate the ideas. The first example shows an interface between the PC and a powerful and popular DSP, the ADSP-2101. The other example shows the interface between the PC and an 8051-like controller, the AT89C2051 microcontroller.

Interface to the ADSP-2101

I needed to connect an ADSP-2101-based Charge Coupled Device (CCD) camera controller to a PC for diagnostic purposes. Eventually the controller would be connected to the PC through a high-speed optical fiber link. The RS-232 interface was ruled out because of the required data transfer rate (>20Kbps). Also, I did not want to invest in a dedicated asynchronous transmission circuit on the controller. The solution was to use the parallel printer adapter on the PC as a link between the camera controller and the PC.

The CCD Camera Controller

The camera controller is built around a fixed-point DSP from Analog Devices, the ADSP-2101. Use of DSPs in embedded control applications reflects a growing trend.

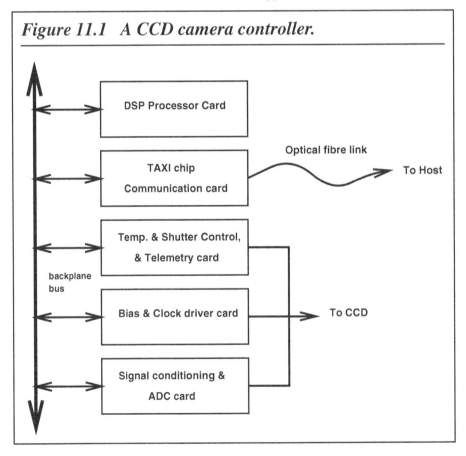

Figure 11.1   A CCD camera controller.

I found DSPs from Analog Devices especially attractive in terms of ease of use, availability, price, development tools, documentation, and customer support.

The ADSP-2101:

- is a fast microcomputer,
- executes any instruction in one clock cycle,
- has zero overhead looping, and
- has two buffered synchronous serial ports capable of transmission rates up to 5Mbps.

As a camera controller, the DSP helps acquire CCD images. The image parameters are set by the user through a host computer. These parameters define the exposure time, size of the CCD image, pixel, row binning, etc. To do this the ADSP-2101:

- receives commands from the host,
- waits for the signal integration time,
- generates CCD clock waveforms to shift out each pixel signal,
- reads CCD pixel signal voltage through an ADC, and
- transfers the ADC data over a suitable link to the host.

Figure 11.1 shows the block diagram for such a camera controller. The serial link between the host and the controller is implemented with a high-speed optical fiber link (in the final configuration). The components of the controller are:

- A backplane bus that carries interconnections between the various cards of the CCD controller.

- An ADSP-2101 processor card to implement a programmable waveform generator, an optional host communication link (using the synchronous serial interface of the DSP), a serial ADC interface, and a backplane bus interface to connect to the other components of the controller. The waveform generator is a crucial component of a CCD controller. Having a programmable waveform generator allows the user to operate the CCD camera in a wide variety of modes by merely downloading a new waveform description table from the host.

- A high-speed (50Mbps) serial link using a TAXI chip set. The TAXI chip set interfaces to a 850nm fiber optic physical link. This card connects to the backplane bus. The TAXI chip receives 8-bit-wide characters from the DSP card, to be transmitted on the fiber link to the host. Received characters from the host are read by the DSP, eight bits at a time.

- A temperature controller, shutter driver, and telemetry card connected to the backplane bus. This card has a temperature controller (of the proportional control type) to maintain the temperature of the CCD chip. The temperature is set by a preset potentiometer on this card. A voltage multiplier using a high-frequency transformer charges a reservoir capacitor. This capacitor discharges into the shutter through a FET switch when the shutter is to be operated. The DSP controls the voltage multiplier and the shutter operation. A multichannel 12-bit serial ADC is used to read chip temperature, dewar temperature, shutter status, etc. This card also has stepper motor drivers for controlling two stepper motors.

- An analog signal processor (with double-correlated sampling) and serial ADC card. This card also connects to the backplane bus. The signal processor circuitry receives the CCD pixel voltage from the backplane bus. This voltage is encoded by the 16-bit serial ADC. The serial ADC control signals on the backplane bus are derived from the serial port of the DSP.

- A CCD clock bias and driver card. This card uses the majority of waveform signals (generated by the DSP processor card) present on the backplane bus. These signals pass through appropriate line drivers before being filtered on the backplane card. The CCD clock levels are referenced by the bias voltage generator on this card. The bias voltages are generated by 24 DACs, which are initialized and controlled by the DSP.

CCD Controller Testing

The most important components of the controller are the DSP processor card, the bias and clock driver card, and the signal conditioning and ADC card. Any attempt at prototyping a CCD controller must start with a focus on these functions. However, in order to test the functions and the performance of the CCD controller, a host computer must send parameters and receive encoded data. A possible solution is to implement an RS-232 serial link to the PC. I ruled out the RS-232 port in this case, however, because of its low data transfer rates and the required investment in additional hardware. Using the parallel printer adapter for bidirectional data transfer would not only give me a high enough data transfer rate, but it would allow me to use common latches, buffers, and decoders. At a software level, only the device-level routines would need to be modified.

A host interface port to connect the controller hardware to the PC through the parallel printer adapter (as described below) simplifies evaluation of ADC performance and allows you to test the waveform generator algorithm and the noise characteristics of the signal-conditioning circuit.

Connecting the Controller and the Parallel Adapter

The circuit in Figure 11.2 converts the parallel printer adapter into a simple host interface port. The port can connect virtually any processor to the PC for bidirectional data transfer. A conservative estimate of the data transfer rates between the PC (a 486/66MHz) and the CCD controller is the range of 20–50Kbps. The test program uses routines in Listing 11.1 for communication. Suitable programs are written for the

Figure 11.2   PC interface for ADSP-2101 DSP.

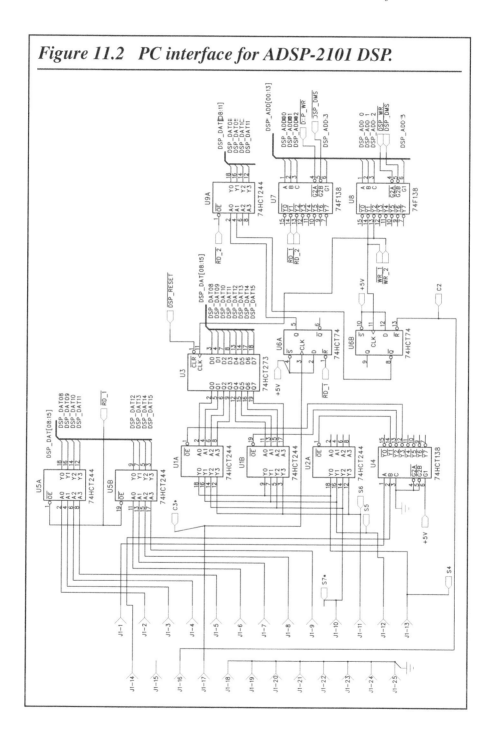

DSP on the CCD controller. Coding parts of the routines in assembler can significantly increase data transfer rates.

To convert the parallel printer adapter into a host interface, the host uses the DATA port to transmit eight bits of data to the application. A flip-flop U6-A is reset to indicate to the application hardware that a data byte is available. This flag could also be

---

Listing 11.1 Connect the PC to the DSP-based CCD controller.

```c
/*dsp_pc.c*/
/* Program to connect the PC through the printer adapter to the
DSP based CCD controller.*/

#include<stdio.h>
#include<conio.h>
#include<dos.h>
#include<process.h>
#include<time.h>

/* external variables */

extern unsigned dport, sport, cport;

/* external routines. gets the addresses of the 3 ports from the DOS

data RAM   */
extern void set_lpt_base_address(int);

/* status port */

#define pin_11      0x80
#define pin_10      0x40
#define pin_12      0x20
#define pin_13      0x10
#define pin_32      0x08

/* control port */

#define pin_17      0x08
#define pin_16      0x04
#define pin_14      0x02
#define pin_1       0x01

/* op & ip semaphores */
#define ip_buffer_flag 0x04
#define ip_buffer_Flag 0xfb
```

used to interrupt the application hardware if necessary. The DSP application hardware, in this case, reads the DATA bits through its input port buffer U5 and sets up the flag U6-A. The host program monitors U6-A before transmitting a new byte.

To receive a byte from the application hardware, the host monitors flag U6-B. If flag U6-B is reset, it indicates that a byte is ready for the host. Reading this byte is a tricky process because the parallel printer adapter is capable of reading only five bits at a time. To overcome this shortcoming, tristate buffers U1 and U2 are used. U1 allows eight bits at its input and transmits only four of these to the output. Nibble control pins 1G and 2G on U1 and U2 are controlled by the decoder outputs of U4 to determine which four of the 16 possible inputs are connected to the output pins. The four output pins of U1 and U2 are connected to the STATUS port bits.

The host program manipulates the decoder U4 to enable the lower four bits of the incoming byte to reach the STATUS port. The STATUS port is then read and its contents temporarily stored away. The decoder is now manipulated to read the upper four

---

Listing 11.1 (continued)

```
/* this flag is on bit 2 (pin 16 ) of the control port
and can set by a logic low on the pin 16*/

#define op_latch_flag 0x08
#define op_latch_Flag 0xf7

/* this flag is set by pulsing a low on pin 17 (bit 3)
 of the control port. SET condition of this flag indicates
 that the oplatch contains a new byte  */

/* local routines */
unsigned char txstat(void);
/* check to see if the o/p latch is empty; empty=0 */

unsigned char rxstat(void);

/* check to see if the i/p buffer has any char; empty=0 */

void tx(unsigned char);     /* transmit the char to the latch */
unsigned char  rx(void);    /* receive a char from the buffer */
void enable_nibble(unsigned char);
/* this function controls which nibble gets connected to
 the status port pins */

/* txstat: This routines checks pin_13 of printer adapter status port
 if the PIN is SET, the o/p latch is full & should not be written to
 again. When the DSP reads the latch, the PIN is RESET. Now the latch
 can be written to again */
```

bits of the byte into the STATUS port. The actual byte is reconstructed by shifting the first STATUS port read four bits to the right and bitwise ORing with the second STA-TUS port read result. The seventh and the third bits are complemented to reconstruct the actual byte. Thereafter, flag U6-B is set to indicate to the application hardware that the current byte has been read by the host. A note of caution here. The process of

---

Listing 11.1 (continued)

```
/* return value: 1 is latch full
   0 if latch empty
*/

unsigned char txstat(void)
{
   char latch_status;
   enable_nibble(1);     /* this function connects the sport to nibble 1*/
   latch_status=inportb(sport) & pin_13;
   return latch_status;
}

/* rxstat: this function checks pin_12 of the status port. If the PIN is
 set, the buffer is full & should be read. if RESET, it is empty. */

/* return value:    0 if the buffer is empty
   1 if the buffer is full
*/

unsigned char rxstat(void)
{
   char buffer_status;
   enable_nibble(1); /* this function connects the sport to nibble 1*/
   buffer_status=inportb(sport) & pin_12;
   return buffer_status;
}

/* tx: This routine latches a byte into the o/p latch */
/* return value: none */

void tx(unsigned char op_byte)
{
   unsigned char temp;
   outportb(dport, op_byte);  /* latch the byte*/

   /*
    now set up the op_latch_flag to indicate that a new byte is
    available
    */
```

reading the eight bits of incoming data and the setting up of the flag by the host is not an atomic operation. The flag is set up by executing a few instructions by the host program. On my 486/66MHz PC this translates to about 5µs.

To have an error-free data transmission, the application hardware waits for, say, 10µs after it detects that the byte has been read by the host before it transmits a new byte to the host. Such a precaution is not required for the data transmission from the host to the application hardware because reading the byte and setting up flag U6-A is an atomic operation.

Listing 11.1 (continued)

```
    temp=inportb(cport) & (0xff ^ op_latch_flag);
    temp=temp ^ op_latch_flag;
    outportb(cport, temp);
    temp=temp ^ op_latch_flag;
    temp=temp | op_latch_flag;
    temp=temp ^ op_latch_flag;
    outportb(cport, temp);
    return;
}
/* rx: This routine reads the i/p 8 bit buffer */
/* return value: the byte read from the buffer */

unsigned char rx(void)
{
    unsigned char ip_byte, temp;
    enable_nibble(3); /* set the buffer to read the lower nibble */
    temp=inportb(sport);
    temp=temp >> 4;
    enable_nibble(2); /* set up the buffer to read upper nibble */
    ip_byte=inportb(sport);
    ip_byte = ip_byte & 0xf0; /* reset lower 4 bits */
    ip_byte=0x88 ^ (ip_byte | temp);

    /* concatenate the 2 nibbles & flip 7th & 3rd bit */
    /* now reset the flag to indicate that the byte has been read */
    temp=inportb(cport) & (0xff ^ ip_buffer_flag);
    outportb(cport, temp);
    temp = temp | ip_buffer_flag;
    outportb(cport, temp);
    return ip_byte; /* return the converted byte */
}
```

Driver Software

A sample program to communicate bidirectionally through the Host Interface Port (HIP) is illustrated in Listing 11.1. The listing contains four routines that help to regulate data flow.

Listing 11.1 (continued)

```
void enable_nibble(unsigned char nibble_number)
{
    unsigned char cport_status;
    cport_status=( inportb(cport) & 0xfc) ;   /* isolate bit 0 & 1*/
    nibble_number = nibble_number & 0x03;
    nibble_number = 0x03 ^ nibble_number;   /* invert bit 0 & 1 */
    cport_status=cport_status | nibble_number;
    outportb(cport, cport_status);
    return;
}
main()
{
    unsigned long count;
    unsigned char portval, tempp, tempq;
    time_t t1,t2;
    FILE *fp1;
    int temp=1;
    clrscr();
    printf("\n\nFinding Printer adapter lpt%d...", temp);
    set_lpt_base_address(temp);
    if(dport == 0)
    {
        printf("\nPrinter adapter lpt%d not installed...", temp); exit(0);
    }
    else
    {
        printf("found. Base address: %xhex", dport);
        portval=inportb(sport);
        printf("\n\n                    D7  D6  D5  D4  D3  D2  D1  D0");
        printf("\nStatus  port value = %x   %x   %x   %x   %x   X   X   X ", \
            (portval & pin_11)>>7, (portval & pin_10)>>6, (portval & pin_12)>>5, \
            (portval & pin_13)>>4, (portval & pin_32)>>3 );
        portval=inportb(cport);
        printf("\nControl port value = X   X   X   X   %X   %X   %X   %X ", \
            (portval & pin_17)>>3, (portval & pin_16)>>2, (portval & pin_14)>>1, \
            (portval & pin_1) );
        portval=inportb(dport);
```

The function `rxstat()` reads the status of flag U6-B, which is reset by the controller to indicate that a byte has been latched into U3. `rxstat()` is used by `main()` to detect the presence of a new byte before reading the byte.

Function `txstat()` returns the status of flag U6-A. When the host program transmits a byte to the controller, it resets th flag. After the controller reads the byte, U6-A ls set by the controller. `txstat()` is used by the host program to ensure that any previously transmitted byte has been read by the controller.

The functions `rx()` and `tx()` receive and transmit a byte, respectively. Function `rx()` reads the byte sent by the controller (as indicated by a logic 0 of flag U6-B). After the byte is read, `rx()` sets U6-B to logic 1. Function `tx()` sets up flag U6-A after a byte is latched into the DATA port register of the printer adapter.

Listing 11.1 *(continued)*

```
   ("\nData    port value = %X   %X   %X   %X   %X   %X   %X   %X ", \
      (portval & 0x80)>>7, (portval & 0x40)>>6, (portval & 0x20)>>5, \
      (portval & 0x10)>>4, (portval & 0x08)>>3, (portval & 0x04)>>2, \
      (portval & 0x02)>>1, portval & 0x01  );
   printf("\n\n\n");
}
/* set up reset states on the control port, all pins to logic 1 */
outportb(cport,0x04);
fp1=fopen("tx_rx", "w");
t1=time(NULL); /* just to log time */
for (count=0;count<256;count++)
{
   while (!txstat());
   /* wait till the DSP application reads the previous byte*/
   tx(tempp); /* transmit a byte*/
   while(rxstat());
   /* wait till a byte is transmitted by the DSP */

   tempq=rx();
   /* byte is available, read it */

   fprintf(fp1, "TX=%x, RX=%x\n ", tempp, tempq);
   /* store it in a file */

   tempp=tempp++;
}
fclose(fp1);
t2=time(NULL);
printf("time taken = %ld secs", t2-t1);
}
```

The main program reads the status of the printer adapter ports and then transmits a sequence of 256 bytes to the DSP application circuit. The application circuit echoes back the received bytes. This is a good test to see if any bytes are missing. I have run this program extensively and found it to work faithfully.

It is possible to add interrupt capability to the HIP. The interrupt signal is derived from pin 10 of the 25-pin D as well as the 36-pin Centronics connector (connected to bit 6 of the STATUS port). A logic 1 on CONTROL port bit 4 enables pin 10 to interrupt the PC. The interrupt signal could be derived from flag U6-A or U6-B. Apart from incorporating an interrupt routine, the current HIP design would need modifications.

It is important to note that with increasing PC speeds, it is quite possible to induce glitches on the CONTROL port bits, especially if you are using the standard printer cable. It is advisable to use a cable with twisted pair wires for each signal.

Interface to the AT89C2051

Atmel's AT89Cxx51 represents a range of high-performance 8-bit microcomputers. AT89C2051 is a low-voltage CMOS microcomputer with 2Kb of flash programmable and erasable read-only memory (PEROM). Components of the AT89Cxx51 series exhibit compatibility with the MSC-51 components with regard to the instruction set and object code. AT89C2051 is manufactured with static internal memory and operates at up to 24MHz clock frequency.

The use of the AT89C2051 is especially attractive for the large number of 8051 devotees because it offers a complete port from the existing 8051 or similar devices in a variety of applications, resulting in reduced board space, components, and cost. Consider a typical system comprised of a suitable EPROM, address latch, 8051, and other associated components. In many applications the 8051 offers enough on-chip digital I/O that additional external I/O components are not required. In such a system, a single AT89C2051 can be used to replace the EPROM, latch, and 8051, provided the system software can be contained within 2Kb, all without investing any time or effort in learning a new system. AT89C2051 is packaged in a 20-pin PDIP as well as SOIC. The AT89C2051 has the following features:

- 2Kb of on-chip FLASH PROGRAM memory
- 128 bytes of internal RAM
- Fully static operation: 0–24MHz
- Instruction compatibility with MCS51
- 15 I/O lines
- Full-duplex programmable serial port
- 16-bit programmable timers
- On-chip analog comparator
- Low-power and power-down modes

- Wide operating voltages: 2.7–6V
- 20-pin DIP/SOIC package

The use of static memory allows the device to be operated down to zero frequency. The AT89C2051 also offers two software-selectable save-power modes. Idle mode stops the CPU, retaining the contents of the internal RAM while the timer/counter, interrupt system, and the serial ports function normally. Power-down mode saves the RAM contents but freezes the oscillator, disabling all other activity until the next hardware reset. Atmel has very judiciously squeezed a host of 8051 hardware features in a 20-pin package, and that is what makes AT89C2051 such an exciting device; coupled with this is the internal 2Kb PEROM, which greatly simplifies the task of reprogramming the device.

The amount of PEROM available on the AT89C2051 is sufficient for a lot of applications, including use in portable instruments, supervisory control applications, autonomous robots, and many more. Use as controllers in portable instruments is further simplified by the low power consumption and wide operating voltage range.

Hardware I/O Features

The AT89C2051 allows 15 bits of I/O configured as eight bits on Port1 and seven bits on Port3. Port1 and Port3 are compatible with P1 and P3 on a 8051, except Port1.0 and Port1.1 .

Port1 pins P1.0 and P1.1 require external pull-ups. P1.0 and P1.1 pins also serve as inputs to an on-chip analog comparator (+ve and –ve inputs, respectively). Port1 output buffers have a 20mA sink current capacity and can drive LEDs directly. By writing 1s to the Port1 bits, they can be used as input bits.

Port3 pins P3.0 to P3.5 and P3.7 are seven bidirectional I/O pins with internal pull-ups. P3.6 is internally connected to the output of the on-chip comparator and is not accessible as a general-purpose I/O pin. Port3 bits can also sink up to 20mA of current and when written with 1s, can be used as inputs.

Port3 pins can also serve the alternative functions listed in Table 11.1. If the user wants to implement these alternative functions, the pin cannot also be used for general-purpose I/O.

Oscillator Characteristics

The AT89C2051 data sheet states that the on-chip oscillator can be used together with a ceramic resonator (as well as a resonant crystal element) to provide the basic clock to the microcomputer. An external clock source with suitable levels can be used instead of a crystal or a resonator. The operation is similar to that of an 8051. AT89C2051 can be operated with a clock frequency between 0 and 24MHz. This is possible because the chip uses static memory.

Special Function Registers

The AT89C2051 has a register set identical to the 8051. Thus it is possible to port any existing 8051 application to an AT89C2051 without changing the object code as long as the software limits itself to the available hardware resources, including memory and ports. This means that all jumps (`ljmp`) and calls (`lcall`) must be limited to a maximum physical address of `0x7FFh`. This also applies to all the other instructions that access memory in some form (e.g., `cjne jmp A + DPTR jnb`, etc). The processor does not support external DATA or PROGRAM memory access.

Using the AT89C2051

Applications that need low current consumption, power-down or sleep modes of operation, a serial port, timers, interrupts, and I/O pins and have small board space could use the AT89Cxx51 series.

I needed a 12-bit multichannel ADC to connect to old, discarded PCs in a certain application. Rather than invest in new parallel ADCs, I decided to see if my existing inventory of MAX186 ADCs could be of any use. The MAX186 had everything I needed, except that it operates at serial clock with a minimum clock frequency requirement of 10µs, which would be difficult to generate under program control on old, slow PCs. I decided to build a general-purpose interface that could be used in other applications. I found that I could use the AT89C2051 very nicely. The result of my design was an elegant solution that offers a nibble-wide input and output interface, which, though tailored to connect to the PC parallel port, can be used anywhere else.

Table 11.1   Port3 alternative functions.

Port Pin	Alternative Function
P3.0	RXD (serial input port)
P3.1	TXD (serial output port)
P3.2	$\overline{\text{INT0}}$ (external interrupt 0)
P3.3	$\overline{\text{INT1}}$ (external interrupt 1)
P3.4	T0 (timer 0 external input)
P3.5	T1 (timer 1 external input)

Parallelizing Serial ADC Data

Using an 8051-capacity controller to parallelize serial ADC data could be considered overkill, but rather than spend time selecting new components, I found it more appropriate to use available components and get the larger instrument working. The AT89C2051 microcontroller uses its serial port signal pins TxD and RxD to connect to the MAX186. The microcontroller serial port operates in mode 0, in which the serial port works as a shift register, either for input or output. In the shift register mode, the TxD pin supplies the shift clock, and the RxD pin provides the data or reads the external data as per the direction. The controller program programs the serial port as an output shift register in the beginning of the acquisition cycle, during which the MAX186 needs 8-bit conversion parameters, channel numbers, etc. After this 8-bit data is shifted out, the controller program converts the serial port as an input shift register and reads back the converted ADC data as two bytes.

Figure 11.3 shows the block diagram and Figure 11.4 shows the circuit schematic of the AT89C2051-to-MAX186 interface connected to the parallel adapter. Figure 11.5 shows the wiring scheme to connect the AT89C2051 controller board to the PC printer adapter. The user interface of the converter consists of the following signals:

- four bits of *mode* inputs that determine the mode of operation for the converter;

- a *trigger* input that triggers the converter into the requested mode; and

- a *clear status* input that is used to erase previous status information.

The outputs of the converter are:

- four bits of *data*;

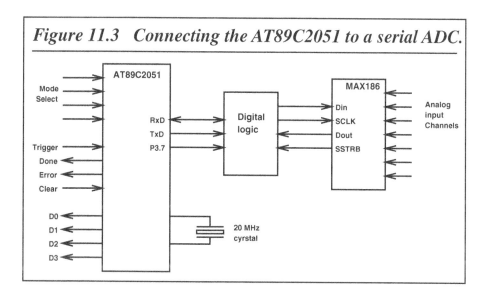

Figure 11.3   Connecting the AT89C2051 to a serial ADC.

Figure 11.4 A parallel adapter interface to connect to a MAX186 serial ADC through the AT89C2051 microcontroller.

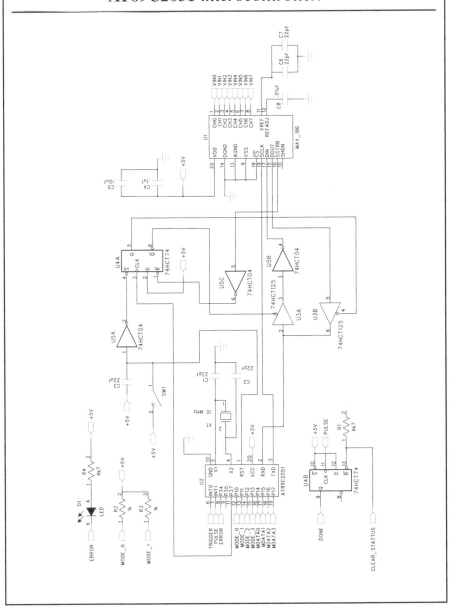

- a *done* flag that indicates the end of last operation; and
- an *error* flag to indicate an attempt to use an unimplemented mode of operation.

The mode input to the converter determines what task the controller will perform when it is triggered. With four bits of mode input, up to 16 modes of operation can be implemented. For this design, only 11 combinations are required. The rest can be used for later expansion. The modes are shown in Table 11.2.

Using the Converter

This interface is designed in such a fashion that it can be used in any embedded application as part of a small data acquisition system or a bigger instrument. It is useful to note that the interface is ideally suited for data acquisition on PC compatibles using the parallel printer adapter. The converter provides access to eight channels of 12-bit ADC. The analog input voltage range of the ADC is 0–4.095V and, at 12 bits, a resolution of 1mV. Listing 11.2 provides the C code to interface the AT89C2051 controller through the PC parallel port.

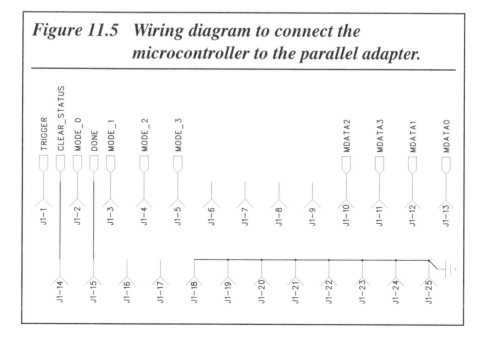

Figure 11.5 Wiring diagram to connect the microcontroller to the parallel adapter.

Table 11.2   Converter operation modes.

Mode	Action
0	start ADC conversion on channel 0
1	start ADC conversion on channel 1
2	start ADC conversion on channel 2
3	start ADC conversion on channel 3
4	start ADC conversion on channel 4
5	start ADC conversion on channel 5
6	start ADC conversion on channel 6
7	start ADC conversion on channel 7
8	read nibble 0 of the last ADC conversion result
9	read nibble 1 of the last ADC conversion result
10	read nibble 2 of the last ADC conversion result

Listing 11.2   Implement the AT89C2051 interface.

```
/*sadc_ip.c*/
#include <stdio.h>
#include <dos.h>
#include <conio.h>
#include <process.h>
#include<time.h>

void main(void)
{
   int dport_lpt1, cport_lpt1, sport_lpt1, del;
   unsigned char adc_status, adc_val, cport, chan_num;
   unsigned char nib_0, nib_1, nib_2, nib_3, temp, nibble[5];

   /* Sign ON */
   clrscr();
   printf("AT89C2051 based serial ADC adapter for the printer port, Version 1.0");
   printf("\nD.V.GADRE");

   /*Get LPT1 port addresses */
   dport_lpt1 = peek(0x40,0x08);
   if(dport_lpt1 == 0)
   {
       printf("\n\n\nLPT! not available... aborting\n\n\n");
```

Listing 11.2 (continued)

```c
    exit(1);
}
printf("\nLPT1 address = %X", dport_lpt1);
cport_lpt1 = dport_lpt1 +2;     /* control port address */
sport_lpt1 = dport_lpt1 + 1;    /* status port address */

/*initialize the ADC strobe signals*/
cport=04;
outportb(cport_lpt1, cport);

/*clear the status of controller*/
/*this generates a low going pulse on C1* pin of the control port*/
cport=cport | 02;
outport(cport_lpt1, cport);
cport=cport & 0xfd;
outport(cport_lpt1, cport);

/* check if ADC is connected & working*/
adc_status=inportb(sport_lpt1);
adc_status=adc_status & 0x08;
if(adc_status == 0) printf("\nADC connected\n");
chan_num=0; /*set channel number*/
for(;;)
{
   /*set channel number to the controller*/
   outportb(dport_lpt1, chan_num);

   /*trigger the controller to read the channel number*/
   cport=cport | 01;
   outportb(cport_lpt1, cport);

   /*wait till it reads the channel number and completes conversion*/
   adc_status=inportb(sport_lpt1);
   adc_status=adc_status & 0x08;
   while(adc_status == 0)
   {
      adc_status=inportb(sport_lpt1);
      adc_status=adc_status & 0x08;
   }

   /*remove the trigger pulse. take it igh to remove the trigger*/
   cport=cport & 0xfe;
   outportb(cport_lpt1, cport);

   /*clear the status of controller*/
   cport=cport | 02;
```

Listing 11.2 (continued)

```
      outport(cport_lpt1, cport);
      cport=cport & 0xfd;
      outport(cport_lpt1, cport);

      /*now read the converted data*/
      for(temp=8; temp<11; temp++)

      {
        /*set data port to read nibble 0, 1 and 2 in that order*/
           outportb(dport_lpt1, temp);

         /*trigger the converter to perform the read nibble process*/
           cport=cport | 01;
           outportb(cport_lpt1, cport);

        /*wait till it completes the task*/
           adc_status=inportb(sport_lpt1);
           adc_status=adc_status & 0x08;
           while(adc_status ==0)
           {
             adc_status=inportb(sport_lpt1);
             adc_status=adc_status & 0x08;
           }

        /*remove the trigger pulse*/
           cport=cport & 0xfe;
           outportb(cport_lpt1, cport);

       /*clear the status of controller*/
           cport=cport | 02;
           outport(cport_lpt1, cport);
           cport=cport & 0xfd;
           outport(cport_lpt1, cport);

       /*read the nibble and store it temporarily*/
           adc_val=inportb(sport_lpt1);
           nibble[temp-8]=(adc_val ^ 0x80) >> 4;
      }

     /*construct the full 12 bit number from the stored nibbles*/
     printf("\n %d mV", nibble[0] + 16*nibble[1] + 256*nibble[2]);
     sleep(1); /*sleep for 1 sec*/
   }
}
```

Chapter 12

Hosting a Device Programmer

This chapter describes how to use the parallel adapter to host a device programmer. In an earlier chapter, you saw how the parallel adapter is used to emulate EPROM/ROM memory. After you have used the emulator to generate the code for a particular microprocessor application, the next step in making the microprocessor circuit run on its own is transferring the developed code into EPROM. To do this, you need an EPROM programmer. If the microcontroller you're using has its own internal memory, you need a programmer that can handle that particular controller. Commercial device programmers can often program a large number of devices ranging from ordinary EPROMs to PAL and PLA devices to microcontrollers. In this chapter I describe an EPROM programmer, as well as a programmer for the AT89C2051 microcontroller that was used in an application in the previous chapter.

An EPROM Programmer

An EPROM is a memory device that allows digital information, which could be a piece of code (as developed with the help of the EPROM emulator), to be stored semi-permanently. The storage of code inside the memory is achieved with the help of special programming algorithms and hardware. After the data is stored in the device, the data is retained, even if the supply voltage is not applied. To erase the data from the chip, the device needs to be exposed to UV radiation for a specified period of time.

Figure 12.1 shows the setup for programming the EPROM. Figure 12.2 shows the timing diagram for programming the EPROM. The timing diagram indicates that the address bus is supplied with the address of the required location that needs to be programmed. Then the program data is applied to the data bus of the EPROM device, and the Vpp pin of the device is applied a voltage of +12.5V. To permanently write the data into the device, a pulse on the CS* pin of the device is applied for 1ms. To ensure that the data has been written properly, the data needs to be read back. To do this, the CS* pin is taken high at the end of the 1ms programming pulse and the OE* pin is taken low. After that the CS* pin is also taken low. This puts the EPROM in read mode and the contents of the location whose address is on the address bus is output on the data bus.

The programmer compares this data with the data that was programmed into the chip. If the comparison is true (that the read-back data is equal to the programmed data), the data write is complete. However, in the case in which the comparison is not true, this procedure is repeated a maximum of 20 times. If the device still is not programmed, the EPROM is faulty and cannot be used. In case the data gets programmed in one of the intermediate cycles, an extra pulse of 1ms duration is applied just as a precaution, and the programming cycle ends. Now the next data byte at the next memory location can be applied. This cycle is repeated until all required data bytes have been programmed into the device.

There are many variants of the actual programming algorithm. In the early days of EPROMs, it was advised to apply a single programming pulse for a 50ms duration.

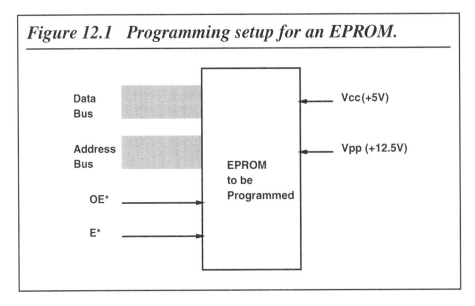

Figure 12.1   Programming setup for an EPROM.

Then a more intelligent approach appeared, in which the supply voltage during programming was raised to 6V and a 100µs programming pulse was applied. For most EPROMs, the method discussed in this section seems to work. If this approach doesn't work for your situation, see the manufacturer's data sheets for a more exact programming approach.

Figure 12.3 shows the suggested circuit diagram for a parallel adapter-based EPROM programmer. The example shows the programmer for 27256-based EPROMs, and it should be easy to adapt the circuit for other EPROMs. Please note that this is not a tested circuit.

The circuit shows four latch ICs (a 74HCT573 and three 74HCT273) connected to the DATA port signals of the parallel adapter. The clock signals of these ICs are provided by the CONTROL port signals C1*, C2, and C3*. The CONTROL port signal that drives the clock input of the 75HCT573 IC is C0*, and the default state for this signal is 0. To latch data into the 74HCT573 latch, the program generates a high-going pulse on C0*. The driver program of the project should put the default state of the respective clock signals on the C0*, C1*, C2, and C3*. For the 74HCT273 octal flip-flops, the PC program generates a low-going pulse on the respective clock input pins.

IC 74HCT573 (U2) supplies the program data to the EPROM. The IC has an output enable control pin, which is controlled by output bit 7 of another latch (IC U1, 74HCT273). At power on, this bit is cleared to 0 and thus the output of the inverter that drives the OE* signal of the 74HCT573 IC is at logic 1, putting the output to a high-impedance (tristate) state. The data bits of the EPROM are read back by the parallel port through the tristate buffers U9A and U9B. When the EPROM data is being

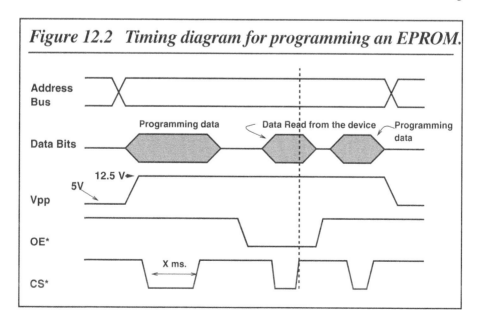

Figure 12.2   Timing diagram for programming an EPROM.

Figure 12.3   A 27256 EPROM programmer.

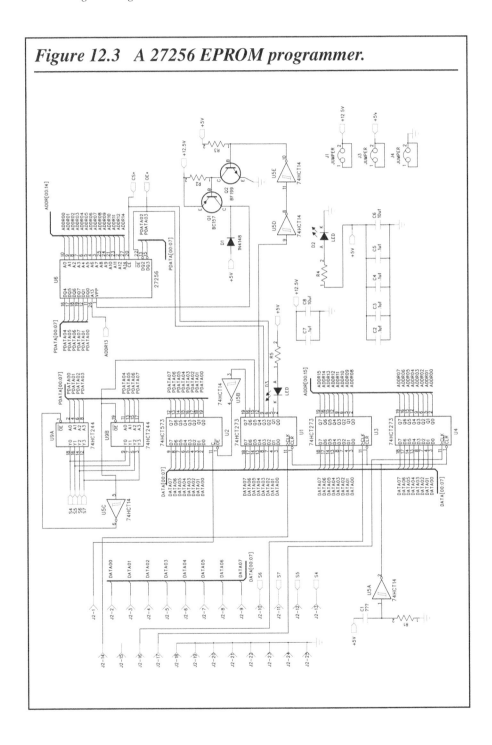

read, the data output of the U2 latch, which is also connected to the EPROM data bits, is disabled under program control. The EPROM outputs internal data when its CS* and OE* signals are low. The OE* and the CS* signals of the EPROM are controlled by bit 3 and bit 2 of the latch IC, U1. At power reset, these are cleared so the EPROM outputs its data on the data pins. The Vpp voltage to the EPROM is supplied by the Q1 and Q2 transistors, which are driven by output bit 4 of the latch IC, U1. At reset, the Q1 transistor is cut off, so the diode D1 conducts, and the Vpp voltage is +5V. When bit 4 of U1 is 1, Q1 conducts and the voltage at the Vpp pin is +12.5V.

The address to the EPROM is supplied by the latch ICs U3 and U4. The address output at reset is 0000h. For every byte, a new address is written into these latches under program control. With 16 address lines, EPROMs up to 64Kb capacity can easily be handled by this circuit. Using the unused lines of latch IC U1, even larger capacity EPROMs can be programmed.

An AT89C2051 Microcontroller Programmer

In the previous chapter, you saw an application based around the 8051-compatible AT89C2051. This microcontroller has internal programmable flash memory. Programming flash memory devices is similar to programming EPROM devices except

Table 12.1 AT89C2051 programming modes.						
Mode	*RST*	*P3.2*	*P3.3*	*P3.4*	*P3.5*	*P3.7*
Write code data[1,2]	12V	low-going pulse	L	H	H	H
Read code data[1]	H	H	L	L	H	H
Write lock bit 1	12V	low-going pulse	H	H	H	H
Write lock bit 2	12V	low-going pulse	H	H	L	L
Chip erase	12V	low-going pulse[3]	H	L	L	L
Read signature byte	H	H	L	L	L	L

[1] Internal PEROM address is reset to 000h on the rising edge of RST and advanced by a positive pulse on the XTAL1 pin.

[2] P3.1 is pulled low during programming.

[3] Chip erase requires a 10ms pulse.

that the required duration of the programming pulse is rather small. In this section, I describe a simple programmer for AT89C2051 built around the parallel adapter.

The AT89C2051 can be programmed using a suitable programmer out of the target system. (The AT89C2051 cannot be programmed in situ.) Table 12.1 shows the various modes for erasing, programming, and verifying the chip.

The code memory is programmed one byte at a time. To reprogram any nonblank byte after the chip has been programmed, the entire chip has to be electrically erased. Erasing the chip is a simple task that requires only a few milliseconds. Figure 12.4 shows the signals required to program the microcontroller chip, and Figure 12.5 shows how to verify the contents of the controller memory.

AT89C2051 Programmer Hardware

Figure 12.6 shows a simple erase–program–verify type of programmer for the Atmel microcontroller. You can add additional features, such as support for security bytes, by modifying the driver software.

The circuit diagram in Figure 12.6 shows a 20-pin ZIF socket into which the controller will be inserted. The controller needs a certain power-up sequence. After the

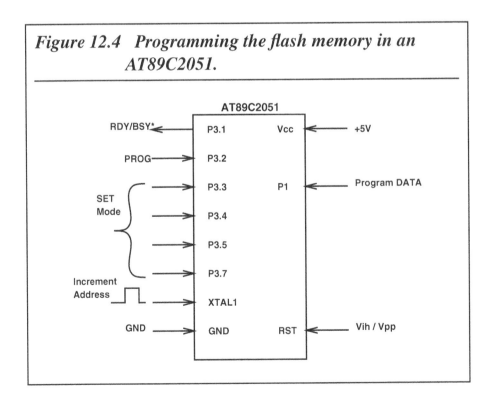

Figure 12.4 Programming the flash memory in an AT89C2051.

power is applied to the controller, the controller must be erased. Only after the controller is erased can it be programmed with the user code.

The program data to the controller is supplied by the output of the 74HCT573 latch IC, U2. The data from the controller can be read back using the U9 buffer IC. When the programmer driver software needs to read the controller data, it disables the U2 latch (that supplies the program data to the controller) and generates signals to the controller to read the data. The 8-bit data is read back from the controller in two nibbles.

The other flip-flop ICs, U1, U3, and U4 (all 74HCT273) are used to generate programming control signals for the controller. Z2 and Z3 are two reed relay switches that are used to switch the power and programming voltage to the controller. On reset, the relays are off and the power to the controller ZIF socket is disconnected. All the other control signals to the socket are in tristate condition.

After reset, the PC program reads the intelhex object code file. It then asks the user to put the controller chip, which needs to be programmed, into the ZIF socket. After the user puts in the chip, the program applies power to the chip and then erases the controller chip.

After the chip has been erased, the PC program applies the necessary program signals to the chip and programs a byte in the first location. These signals include the

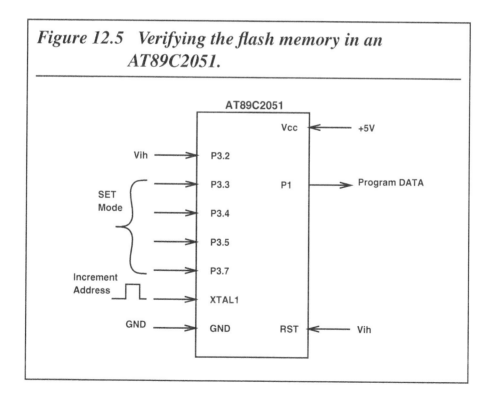

Figure 12.5 Verifying the flash memory in an AT89C2051.

Figure 12.6   Atmel AT89C2051 programmer.

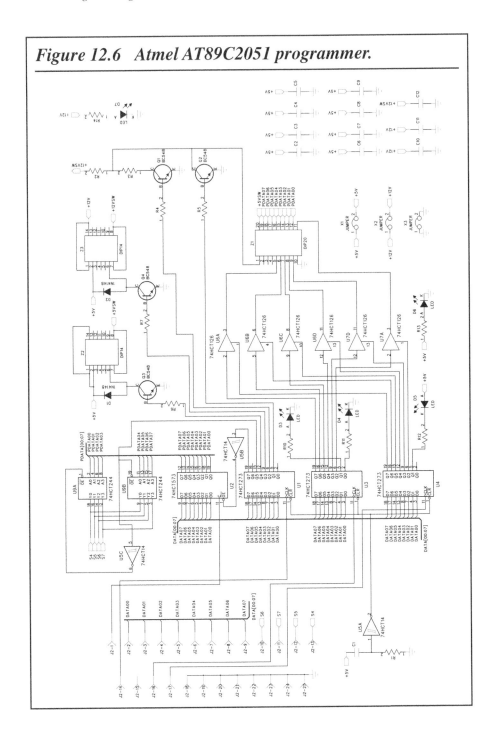

mode selection bits on the P3.3, P3.4, P3.5, and P3.7 bits of the controller chip, as described in the last section. To program the byte, the PC applies a +12V pulse using the voltage switch comprised of transistors Q1 and Q2. It then applies control signals to read back the programmed byte from the controller. This read-back byte is compared with the original byte to check whether the controller is being programmed correctly. In case of failure, the PC program prints an error message, disconnects all power and control signals from the ZIF socket, and exits. If the read-back byte compares correctly with the programmed byte, the PC generates control signals to increment the internal address counter of the controller chip (by generating a pulse on the XTAL1 pin) and programs subsequent bytes into the controller until all 2,048 bytes have been programmed into the controller.

Listing 12.1 shows the AT89C2051 programmer driver software.

Listing 12.1 AT89C2051 programmer driver software.

```
/*
parx.c
Atmel AT89C2051 Programmer
Dhananjay V. Gadre
*/

#include<stdio.h>
#include<conio.h>
#include<dos.h>
#include<process.h>
#include<time.h>
#include<alloc.h>
#include<ctype.h>
#include<stdlib.h>
#include<string.h>

/*port addresses of the parallel adapter*/
unsigned int dport, sport, cport;

/*these ports control data to the uC, voltage to the ZIF socket,*/
/*control the tri-state buffers respectively*/
```

Listing 12.1 (continued)

```
unsigned char port_0, port_1, port_2, port_3, error_byte;

#define MEMORY 2048 /*last address of the target controller memory*/

/*the Intelhex file has lines of code. each line begins with a : */
/*next is number of bytes in the line*/

#define LL 1      /*offset in the hex file where the line length is stored*/
#define ADDR 3   /*offset in the hex file where the destination address is stored*/
#define ZEROS 7
#define CODE_ST 9 /*offset of the beginning of the code*/

/* status port */
#define pin_11     0x80
#define pin_10     0x40
#define pin_12     0x20
#define pin_13     0x10
#define pin_15     0x08

/* control port */
#define pin_1     0x01
#define pin_14    0x02
#define pin_16    0x04
#define pin_17    0x08


/*define to be used with port_1*/
#define ENB_DATA 0X80     /*OR this*/
#define ENB_LOW 0X20      /*OR this*/
#define ENB_HIGH 0xdf     /*1101 1111, AND this*/
#define SW_12V_ON 0x08    /*OR this*/
#define SW_5V_ON 0x10     /*OR this*/
#define SW_12V_OFF 0xF7   /*AND this*/
#define SW_5V_OFF 0xEF    /*AND this*/
#define PULSE_0 0x06      /*  OR this 0000 0110 */
#define PULSE_5 0xFD      /* AND this 1111 1101 */
#define PULSE_12 0xF9     /* AND this 1111 1001 */


/*define to be used with port_2*/
#define XTAL1 0x80
#define P32    0x40
#define P33    0x20
#define P34    0x10
#define P35    0x08
#define P37    0x04
```

Listing 12.1 (continued)

```c
/*defines to be used with port_3*/
#define XTAL1_CON 0x80
#define P32_CON   0x40
#define P33_CON   0x20
#define P34_CON   0x10
#define P35_CON   0x08
#define P37_CON   0x04


/*local global variables*/
unsigned char  ram[2100];
unsigned int curr_address;
FILE *fp1;

/* local routines */
int initialze(void);   /* initialzes the external hardware */
int fill_buffer(void); /*read the intelhex format file & fill up the internal buffer*/
int chk_programmer(void); /*check if the programmer is connected and if +12V is ON*/
int erase_chip(void);
int burn_verify_bytes(void);
int v0_on(void);  /*apply 0volts on RST pin*/
int v5_on(void);  /*apply 5volts onm the RST pin*/
int v12_on(void); /*apply 12V on the RST pin*/
int power_off(void); /*remove power to the ZIF socket and float all pins*/
int power_on(void);  /* apply power and put 0 V on RST and XTAL1 pin*/
                     /*rest all pins float and wait for more than 10 ms*/
void shutdown(void); /*routine to disable everything and to shutdown power*/
                     /*so that the chip can be removed*/

/*routines to generate pulse on each of the 4 control port pins*/
void pulse_c0(void);
void pulse_c1(void);
void pulse_c2(void);
void pulse_c3(void);

void pulse_c0(void)
{
   unsigned char temp;

   temp=inportb(cport);
   temp=temp & 0xfe;
   outportb(cport, temp);
   delay(1);
   temp=temp | 0x01;
   outportb(cport, temp);
   delay(1);
}
```

Listing 12.1 (continued)

```
void pulse_c1(void)
{
    unsigned char temp;

    temp=inportb(cport);
    temp=temp | 0x02;
    outportb(cport, temp);
    delay(1);
    temp=temp & 0xfd;
    outportb(cport, temp);
    delay(1);
}

void pulse_c2(void)
{
    unsigned char temp;
    temp=inportb(cport);
    temp=temp & 0xfb;
    outportb(cport, temp);
    delay(1);
    temp=temp | 0x04;
    outportb(cport, temp);
    delay(1);
}

void pulse_c3(void)
{
    unsigned char temp;

    temp=inportb(cport);
    temp=temp | 0x08;
    outportb(cport, temp);
    delay(1);
    temp=temp & 0xf7;
    outportb(cport, temp);
    delay(1);
}

char chartoi(char val)
{
    unsigned char temp;
    temp = toupper(val);
        if(temp>0x39) {temp = temp -0x37;}
        else {temp=temp-0x30;}
    return temp;
}
```

Listing 12.1 *(continued)*

```c
int initialize(void)
{
   dport = pook(Qx40, 9);
   sport=dport+1;
   cport=dport+2;
   if(dport ==0) return 0;

   outportb(dport, 0);
   outportb(cport, 0x05); /*all cport outputs high, except C0*/
   outportb(cport, 0x0a); /*all cport pins are low, except C0*/
   outportb(cport, 0x05); /*all cport outputs high, except C0*/
   port_0=0;
   port_1=0;
   port_2=0;
   port_3=0;

   return 1;
}


int fill_buffer(void)

/*read the intelhex format file & fill up the
internal buffer */
{
   unsigned char ch, temp4, temp1, temp2, temp3;
   unsigned char chk_sum=0, buf[600], num[10];
   int line_length, address, line_temp, tempx, count=0;

   count=0;
   while(!feof(fp1) )
   {
     chk_sum=0;

     /* check if start of line = ':' */
     fgets(buf, 600, fp1);
     tempx=strlen(buf);

     /*printf("\n\nString length=%d\n", tempx);*/
     /*printf("\n\n%s", buf);*/

     if( buf[0] != ':')
     {
        printf("\nError... Source file not in Intelhex format. Aborting");
        fclose(fp1);
        return 0;
     }
```

Listing 12.1 (continued)

```c
/* convert the next 2 characters to a byte which equals line length */
temp1=buf[LL];
temp2=buf[LL+1];
if( !isxdigit(temp1) )
{
    printf("\nError in source file. Aborting");
    fclose(fp1);
    return 0;
}
if( !isxdigit(temp2) )
{
    printf("\nError in source file. Aborting");
    fclose(fp1);
    return 0;
}

temp4 = chartoi(temp1);
chk_sum=chk_sum + 16*temp4;
line_length=(int)temp4;

temp4=chartoi(temp2);
chk_sum=chk_sum + temp4;
line_length = 16*line_length + (int)temp4;

/*printf("Entries=%d  ", line_length);*/

if(line_length ==0)
{
    return count;
}
temp1=buf[ADDR];
temp2=buf[ADDR+1];
temp3=buf[ADDR+2];
temp4=buf[ADDR+3];
if( !isxdigit(temp1) )
{
    printf("\nError in source file. Aborting");
    fclose(fp1);
    return 0;
}
if( !isxdigit(temp2) )
{
    printf("\nError in source file. Aborting");
    fclose(fp1);
    return 0;
}
```

Listing 12.1 (continued)

```c
    if( !isxdigit(temp3) )
    {
        printf("\nError in source file. Aborting"),
        fclose(fp1);
        return 0;
    }
    if( !isxdigit(temp4) )
    {
        printf("\nError in source file. Aborting");
        fclose(fp1);
        return 0;
    }
ch = chartoi(temp1);
temp1=ch;

ch = chartoi(temp2);
temp2=ch;

chk_sum = chk_sum + 16*temp1 + temp2;

ch = chartoi(temp3);
temp3=ch;

ch = chartoi(temp4);
temp4=ch;

chk_sum = chk_sum + 16*temp3 + temp4;
address = 0x1000 * (int)temp1 + 0x100 * (int)temp2 + 0x10*(int)temp3
        + (int)temp4;
/*printf("Start Address=%x hex\n", address);*/
if(address > MEMORY)
{
    printf("\nError in source file. Bad address. Aborting");
    fclose(fp1);
    return 0;
}
/*check for the next byte. It has to be 00 **/
temp1=buf[ZEROS];
temp2=buf[ZEROS+1];

if( !isxdigit(temp1) )
{
    printf("\nError in source file. Aborting");
    fclose(fp1);
    return 0;
}
```

Listing 12.1 (continued)

```c
if( !isxdigit(temp2) )
{
    printf("\nError in source file. Aborting");
    fclose(fp1);
    return 0;
}

ch = chartoi(temp1);
temp1=ch;
ch=chartoi(temp2);
temp2=ch;
ch = 16*temp1 + temp2;
if(ch != 0)
{
    printf("\nError... Source file not in Intelhex format. Aborting");
    fclose(fp1);
    return 0;
}

/* now read bytes from the file & put it in buffer*/

for(line_temp=0; line_temp<line_length; line_temp++)
{
    temp1=buf[2*line_temp+CODE_ST];
    temp2=buf[2*line_temp+CODE_ST+1];

    if( !isxdigit(temp1) )
    {
        printf("\nError in source file. Aborting");
        fclose(fp1);
        return 0;
    }

    if( !isxdigit(temp2) )
    {
        printf("\nError in source file. Aborting");
        fclose(fp1);
        return 0;
    }
    ch = chartoi(temp1);
    temp1=ch;
    ch=chartoi(temp2);
    temp2=ch;
    ch = 16*temp1 + temp2;

    chk_sum=chk_sum + ch;
```

Listing 12.1 *(continued)*

```
        if(address > MEMORY)
        {
            printf("\nError in source file. Bad address. Aborting");
            fclose(fp1);
            return 0;
        }
        /*    printf("%X ",ch);*/
        ram[address]=ch;
        address++;
        count++;
    }

    /*get the next byte. this is the chksum */
    temp1=buf[2*line_length+CODE_ST];
    temp2=buf[2*line_length+CODE_ST+1];
    if( !isxdigit(temp1) )
    {
        printf("\nError in source file. Aborting");
        fclose(fp1);
        return 0;
    }
    if( !isxdigit(temp2) )
    {
        printf("\nError in source file. Aborting");
        fclose(fp1);
        return 0;
    }
    ch = chartoi(temp1);
    temp1=ch;
    ch=chartoi(temp2);
    temp2=ch;
    ch = 16*temp1 + temp2;
    chk_sum=chk_sum + ch;

    /*printf("csum=%x", chk_sum);*/
    if(chk_sum !=0)
    {
        printf("\nChecksum Error... Aborting");
        fclose(fp1);
        return 0;
    }
}

return count;
}
```

Listing 12.1 (continued)

```c
int erase_chip(void)
{
   unsigned char tcmp;

   /*to erase chip put sequence of P3.3 P3.4 P3.5 P3.7
                                     1    0    0    0   */

   /*then apply 12V to RST pin*/
   /*then pulse P3.2 low for 10 ms*/
   /*put RST to 5 V again*/

   port_2=port_2 | P33;
   port_2=port_2 & ~(P34);
   port_2=port_2 & ~(P35);
   port_2=port_2 & ~(P37);

   /*prepare the control bits of P3.3, P3.4, P3.5, P3.7*/
   port_3=port_3 | P33_CON | P34_CON | P35_CON | P37_CON;

   outportb(dport, port_2);

   pulse_c2();

   outportb(dport, port_3);
   pulse_c3();

   v12_on(); /*set RST to 12V*/

   /*now generate a pulse on P3.2*/

   port_2=port_2 & ~(P32);
   outportb(dport, port_2);
   pulse_c2();

   delay(100);

   port_2=port_2 | P32;
   outportb(dport, port_2);
   pulse_c2();

   v5_on(); /*put RST to 5 Volts now*/


   return 1;

}
```

Listing 12.1 *(continued)*

```c
int burn_verify_bytes(void)
{
    unsigned char temp, low_temp, high_temp,
    int program_length;
    /*put P3.3 P3.4 P3.5 and P3.7 to 0111*/
    /*apply data to the port0*/
    /*put RST to 12V*/
    /*pulse P3.2 for 2 ms*/
    /*RST to 5V*/
    /*P3.3, P3.4, P3.5 P3.7 to 0011*/
    /*read data back and compare*/

    for(program_length=0; program_length<MEMORY; program_length++)
    /*for(;;)*/
    {
        curr_address=program_length;
        port_2=port_2 & ~(P33);
        port_2=port_2 | P34 | P35 | P37;
        outportb(dport, port_2);
        pulse_c2();

        /*this puts the data into the data latch but the latch is yet to
          be enabled*/
        port_0=ram[program_length];
        outportb(dport, port_0);
        pulse_c0();

        /*put 12V on the RST pin*/
        v12_on();

        /*enable the data on the data pins*/
        port_1=port_1 | ENB_DATA;
        outportb(dport, port_1);
        pulse_c1();
        delay(1);

        /*now generate a pulse on P3.2*/
        port_2=port_2 & ~(P32);
        outportb(dport, port_2);
        pulse_c2();

        delay(10);

        port_2=port_2 | P32;
        outportb(dport, port_2);
        pulse_c2();
```

Listing 12.1 (continued)

```
/*now disable the data latch which sent the programming data*/
port_1=port_1 & ~(ENB_DATA);
outportb(dport, port_1);
pulse_c1();

delay(1);
/*put 5V on the RST pin*/
v5_on();

/*set P3.3 P3.4 P3.5 and P3.7 to read back the data*/
port_2=port_2 & ~(P33) & ~(P34);
port_2=port_2 | P35 | P37;
outportb(dport, port_2);
pulse_c2();

/*read low nibble*/
port_1=port_1 | ENB_LOW;
outportb(dport, port_1);
pulse_c1();
low_temp=inportb(sport);

/*read high nibble*/
port_1=port_1 & ENB_HIGH;
outportb(dport, port_1);
pulse_c1();
high_temp=inportb(sport);

temp= (high_temp & 0xf0) | ( (low_temp >>4) & 0x0f);
temp=temp ^ 0x88;
if(temp != ram[program_length])
{
   error_byte=temp;
   printf(
     "\nError in program verify at address %X (hex). Aborting...",
     program_length);
   printf(
     "\nProgram data %X, read back data %X\n".
     ram[program_length], temp);
   shutdown();
   return 0;
}

/*generate a pulse on XTAL1 to increment the address counter*/
port_2= port_2 | XTAL1;
outportb(dport, port_2);
pulse_c2();
```

Listing 12.1 (continued)

```
      port_2=port_2 & ~(XTAL1);
      outportb(dport, port_2);
      pulse_c2();
   }
   return 1;
}


int v0_on(void)   /*apply 0 volts on RST pin*/
{
   unsigned char port_val, temp;

   port_1=port_1 | PULSE_0; /*PULSE_0 is 0x06*/
   outportb(dport, port_1);

   /*now generate a low pulse on C1* */
   pulse_c1();
   return 1;
}


int v5_on(void)   /*apply 5 volts on the RST pin*/
{
   unsigned char port_val, temp;

   port_1=port_1 | PULSE_0;      /*PULSE_0 is 0000 0110*/
   port_1=port_1 & PULSE_5;      /*PULSE_5 is 1111 1101*/
   outportb(dport, port_1);

   pulse_c1();

   return 1;
}


int v12_on(void) /*apply 12 Volts on the RST pin*/
{
   unsigned char port_val, temp;

   port_1=port_1 | PULSE_0;      /*PULSE_0   is 0000 0110*/
   port_1=port_1 & PULSE_12;     /*PULSE_12 is 1111 1001*/
   outportb(dport, port_1);
   pulse_c1();

   return 1;
}
```

Listing 12.1 (continued)

```
int power_off(void) /*remove power to the ZIF socket and float all pins*/
{
   unsigned char port_val, temp;

   v0_on();                                  /* put 0 volts on the RST pin*/

   port_1=port_1 & SW_12V_OFF;
   port_1=port_1 & SW_5V_OFF;
   outportb(dport, port_1);       /*switch off power to the ZIF socket*/
   pulse_c1();

   return 1;
}

int power_on(void) /* apply power and put 0 V on RST and XTAL1 pin*/
                   /* rest all pins float and wait for more than 10 ms*/
                   /* then put RST and P32 to 5V*/
{
   unsigned char port_val, temp;

   v0_on();                                       /*put 0V on RST pin*/

   port_1=port_1 | SW_12V_ON;                        /*prepare port_1*/
   port_1=port_1 | SW_5V_ON;

   port_3=port_3 | XTAL1_CON; /*enable the XTAL1 control pin of port3*/

   temp=inportb(cport);                         /*prepare pulse on c1*/
   temp=temp | 0x02;

   outportb(dport, port_1);                        /*output for port_1*/

   outportb(cport, temp);            /*pulse c1 low and back to high*/
   temp=temp & 0xfd;
   outportb(cport, temp);

   outportb(dport, port_3);
      /*this puts XTAL1 control to 1 and hence XTAL1 to 0*/
   temp=temp | 0x08;                     /*pulse c3 low and high again*/
   outportb(cport, temp);
   temp=temp & 0xf7;
   outportb(cport, temp);

   delay(2);
   sleep(1);
```

Listing 12.1 (*continued*)

```c
    /*now put RST and P3.2 to 5 Volts*/

    port_2=port_2 | P32,     /*to make P32 high*/
    port_3=port_3 | P32_CON; /*to enable P32 to reach the ZIF pin*/

    v5_on();                 /*this makes RST 5V*/

    outportb(dport, port_2);
    pulse_c2();

    outportb(dport, port_3);
    pulse_c3();

    return 1;
}

void shutdown(void)
{
    port_3=0;
    outportb(dport, port_3);
    pulse_c3();

    port_1=0;
    outportb(dport, port_1);
    pulse_c1();

    port_2=0;
    outportb(dport, port_2);
    pulse_c2();

    return;
}


main(argc, argv)
int argc;
char *argv[];
{
    time_t start, endt;
    unsigned long temp;
    int byte_value, return_val, total_bytes;

    printf("\n\n\n\tAtmel AT89C2051 uC Programmer: Version 1.0\n");
    printf("\t------------------------------------------\n");
    printf("\t\t    Dhananjay V. Gadre");
    printf("\n\t\t      April 1997.\n"); /* 30th April 1997*/
```

Listing 12.1 (continued)

```
if(argc != 2)
{
    printf("\nError... Specify Intelhex source filename. Aborting");
    printf("\nFormat: AtmelP intelhex_sourcefile");
    exit(-1);
}

if((fp1=fopen(argv[1], "r")) == NULL)
{
    printf("\nError...Cannot open source file. Aborting");
    exit(-1);
}

return_val=initialize(); /*Initialize the printer adapter port*/
if(return_val == 0)
{
    printf("\nLPT1 not available. Aborting...");
    fclose(fp1);
    exit(-1);
}
printf("\nLPT1 DATA port address = %X (hex)", dport);

power_off();

printf("\nReading Intelhex source file...:");

return_val=fill_buffer();
if(return_val == 0)
{
    exit(0);
}
printf("\nIntel hex file %s read successful. Total bytes read =%d",
        argv[1], return_val);

fclose(fp1);
printf("\n\nPut the target microcontroller in the ZIF socket and press a key\n");

getch();
power_on();
printf("\nErasing the Microcontroller...\n");
erase_chip();
printf("\nProgramming and Verifying...\n");

return_val=burn_verify_bytes();
```

Listing 12.1 (continued)

```
if(return_val == 0)
{
    printf("\nFailed to program the controller at address: %X (hex)\n",
           curr_address);
    printf("Program value: %X\n", ram[curr_address]);
    printf("Verify value: %X\n", error_byte);
    exit(-1);
}
printf("\nMicrocontroller programmed and verified");
power_off();
shutdown();

printf("\nNow remove the controller from the ZIF socket and press a key");
getch();
}
```

## Chapter 13

Waveform Generation Using the Parallel Adapter

Multichannel digital waveform generators are very useful as general-purpose test instruments for generating known synchronized test waveforms. Another area where multichannel waveform generators are extremely useful is CCD cameras. CCD chips used in these cameras rely on a number of precise, synchronized clock waveforms to collect the charge accumulated in each of the light-sensitive pixels at the output pin of the chip. This chapter looks at some ideas for generating multichannel digital waveforms. After you discover that a table-top instrument with a multichannel, arbitrary digital waveform generation facility doesn't come cheap, you may want to consider some of the following alternatives.

The Parallel Adapter as a Waveform Generator

You can use the output signals of the parallel adapter under suitable program control to generate many channels of digital waveforms. First, plot the required states for the waveform generator on a sheet of paper. Then mark the time duration of each state and assign each state to one of the many signals of the parallel port. Thereafter, bunch the relevant states together and tabulate the numbers representing each state.

Figure 13.1 shows an example of a waveform composed of five digital waves with 12 states. The relative duration of each state is shown in the figure. The minimum duration is 1 unit and the maximum is 4 units. I'll assign 1 unit to be 10μs. Be aware of the limits that this arrangement poses (e.g., it may not be possible to generate this bunch of waveform with 1 unit of 100ns). Typically, waveforms in which each state is ≥1μs may be possible.

The figure is also marked with the logic of each wave. The next step is to assign each of the waves to a bit of a parallel adapter port. I will assign these waveforms to the bits of the DATA port as follows: waveform 0 to D0, 1 to D1, and so on. The unused bits of the DATA port have been set to logic 0. The 12 states (and duration) are as follows:

1. 0x16h, (3 units)

2. 0x1Fh, (3 units)

3. 0x1Eh, (1 unit)

4. 0x14h, (4 units)

5. 0x16h, (1 unit)

6. 0x0Ah, (1 unit)

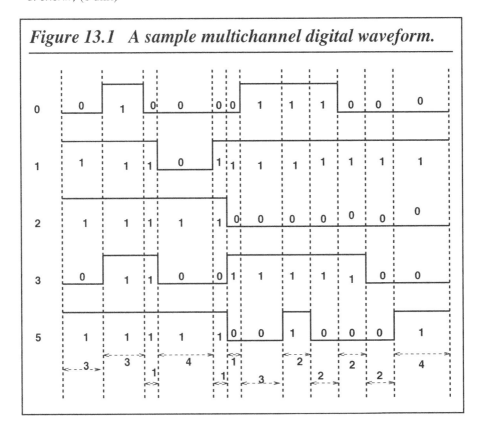

Figure 13.1   A sample multichannel digital waveform.

7. 0x0Bh, (3 units)

8. 0x1Bh, (2 units)

9. 0x0Bh, (2 units)

10. 0x0Ah, (2 units)

11. 0x02h, (2 units)

12. 0x12h, (4 units)

I must now write a program that allocates a buffer of size 12 and initialize the buffer with the values generated above. Next, I need to create a delay routine that provides a delay of 1 unit, which in this case corresponds to 10µs. Listing 13.1 illustrates this example. You should create a delay routine specific to your own hardware if you want to run the waveform generation program. For this case, the unit of each state is 10µs, so the delay routine provides delay in multiples of 10µs, depending upon the argument passed to it. The main program initializes two arrays, one with the required state values and the other with the corresponding state time. Then an infinite loop is entered in which, first of all, the external interrupts are disabled. This ensures that during the time the waveform generator is generating the required states, it runs uninterrupted. At the end of one run, the interrupts are enabled so that pending interrupts can be serviced before the waveforms are generated again.

Listing 13.1   *Arbitrary waveform generator.*

```
/*sarbit.c*/
/*Super simple Arbitrary waveform generator*/

#include<stdio.h>
#include<dos.h>

#define MAX_STATES 12

#define UNIT 100 /*define the count for UNIT*/
/*for our case UNIT must translate to 10 us*/

/*a specific delay routine tuned to produced the required delay*/

void mydel(unsigned int);

void mydel(unsigned int dtime)
{
   unsigned int looptime, temp;
   for(looptime=0; looptime<dtime; looptime++)
     for(temp=0;temp<UNIT;temp++);
}
```

As you can see, this scheme has many problems. For every waveform generation requirement, you have to plot the waveform and mark all state and time entries. Then you have to alter Listing 13.1 to meet specific requirements. Also, you will have to adjust the delay count if you want to use the program on a different machine.

The other serious problem is the lack of repeatability of the waveforms. After each bunch of waveforms is generated, the system interrupts are enabled — you cannot disable the interrupts forever — which means that the time after which the waveform generator will run again is arbitrary; for serious applications this may be unacceptable.

Traditional Methods of Waveform Generation

Conventional methods of digital waveform generation involve storing the required waveform patterns in memory and then clocking a binary counter that provides address and readout signals to the memory device. The diagram in Figure 13.2 illustrates this scheme. The memory device, such as an EPROM or ROM is loaded with the required waveform pattern. The unit of time is the time period of the basic clock source, which in turn is dictated by the required resolution. If the two closest edges

---

Listing 13.1 (continued)

```
main()
{   unsigned char wstate[12], temp=0;
    unsigned int wdelay[12];
    unsigned int dport=0x378;
    /*set the address of the DATA port*/

    /*assign the required states to the array elements*/
    wstate[0]=0x16;
    wstate[1]=0x1F;
    wstate[2]=0x1E;
    wstate[3]=0x14;
    wstate[4]=0x16;
    wstate[5]=0x0A;
    wstate[6]=0x0B;
    wstate[7]=0x1B;
    wstate[8]=0x0B;
    wstate[9]=0x0A;
    wstate[10]=0x02;
    wstate[11]=0x12;

    /*assign the required delay time to the array elements*/
    wdelay[0]=3;
    wdelay[1]=3;
```

are 1μs apart, the basic clock of the system must be at least 1μs. If the total duration of the sequence is 100ms, the number of memory locations would be 100,000 locations. So for an 8-bit digital waveform with a resolution of 1μs and a length of 100ms, a 128Kb EPROM would be suitable for storing the required sequence.

Simpler methods of waveform generation using this scheme would probably store the required waveform at one time outside the waveform generation circuit (using, say, an EPROM programmer), then the memory chip would be plugged into the circuit. For in situ waveform loading capability, a more complex control circuit (along with a communication link to a host) would be needed.

One drawback of this scheme is that it requires a large memory capacity. That in itself is not a problem, but one has to generate large sequences of data and carefully program the memory chip. Any error in this would result as an error in the output waveform. With so many bytes to program, the chances of mistake are not insignificant. The total number of bytes required to store the waveform sequence is equal to the ratio of the total time of the waveform to the required resolution. (In the last example, the ratio r is 100ms/1μs = 100,000 bytes.) If more digital signals are required, additional memory is required. This scheme of waveform generation would need 28

Listing 13.1 (continued)

```
wdelay[2]=1;
wdelay[3]=4;
wdelay[4]=1;
wdelay[5]=1;
wdelay[6]=3;
wdelay[7]=2;
wdelay[8]=2;
wdelay[9]=2;
wdelay[10]=2;
wdelay[11]=4;
/*do it for ever*/
for(;;)
{
      disable(); /*disable external interrupts*/
      /*generate the waveforms with getting disturbed*/
      for(temp=0; temp<MAX_STATES; temp++)
      {
          outportb(dport, wstate[temp]);
          mydel(wdelay[temp]);
      }
      enable(); /*now enable interrupts again*/
      /*so that pending interrupts can be serviced*/
  }
}
```

memory locations. The clock to increment the binary counters generating the memory address runs at 1µs (because this is the minimum time between any two edges). If, however, I alter the waveform just a little bit so that the time between two of the edges is 100ns, the number of required memory locations increases to 280 locations.

An Unconventional Method of Waveform Generation

In addition to the common method of generating digital waveform sequences discussed in the preceding section, another method can save on the required number of memory bytes. The previous method encodes the required state in memory locations. The alternative method not only stores the required state but also stores the required duration for that state. Thus, instead of storing each and every entry at a rate deter-

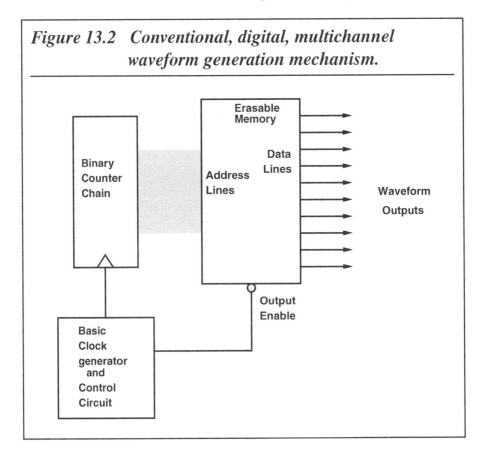

Figure 13.2 Conventional, digital, multichannel waveform generation mechanism.

mined by the required resolution for the required waveform table, this method only stores the transitions and the time between two transitions. The required time between each transition is loaded into a programmable down-counter, which is clocked by a source of clock at a rate equal to the required minimum time on the waveform sequence. With every clock tick, the counter decrements by one, and when the count reaches zero, the counter stops. This generates the timer expire signal. The timer expire signal enables the waveform state FIFO and reads the top byte, latching it into the output state latch. The timer expire signal also enables the lower waveform duration FIFO, reads out the entry, and loads it into the counter. The counter then starts to count down again till it reaches zero, and the output of the state latch remains stable for this duration.

The FIFOs are loaded by the PC with suitable values, as required by the particular waveform sequence. These values are previously calculated and stored as a waveform table file, say, in a simple four-digit hex format, as shown in Listing 13.2. The first two hex digits may represent the state and the next two hex digits the duration. If I use a programmable timer with 8-bit input, the duration could be changed between 1 and 255 units of clock period. For some cases, a duration of 255 units may be insufficient, so a repeat entry could also be used, as shown in Listing 13.2.

The reason for using the FIFOs to store waveform state and duration values is that as long as the FIFOs are not empty, waveform generation can go on without any active supervision of the PC. This means that precise waveform sequences can be generated. The PC monitors the FIFO-full signal and, as long as the FIFO is not full (or the end of the waveform table file is not reached), it transfers data to the two FIFOs. The depth of the FIFOs should be sufficient to tide over any period for which

---

Listing 13.2   Structure of a file to store the waveform table.

```
/*Possible structure of the file to store the waveform table*/
/*The file has 10 entries*/

1603 /*1st entry*/
2340
2060
21FF /*here the duration required is more than the 8-bit range of
2180 /*the time. So a repeat entry is used in this line*/
2050
1021
1F02
0F20
0E40

/*end of the file*/
```

the PC cannot service the FIFOs. If the worst case latency in writing to the FIFOs is 100µs (due to servicing interrupts or other activities) and if the frequency of the basic clock is 100ns, with an assumption that a worst case waveform is only 1 unit of time for every state, the FIFO should have sufficient depth that it can survive 100µs. With 100ns for each waveform state, the required FIFO depth is 1,000. Thus a 1Kb-deep FIFO can be used.

You can see that this method is much like the first method of waveform generation (which used the ports of the parallel adapter to output the required sequence under program control), except here the time between the two transitions of the waveform sequence is counted by an external timer.

The block diagram of this unconventional waveform generator is shown in Figure 13.3. For an 8-bit implementation, two FIFOs, each of 8-bit width, are required. A suitable programmable timer could be the 74LS592, which is an 8-bit programmable up-counter.

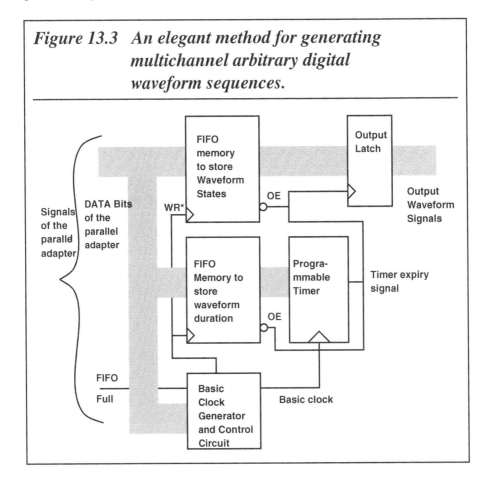

Figure 13.3   *An elegant method for generating multichannel arbitrary digital waveform sequences.*

Chapter 14

Data Acquisition Under Linux

This chapter describes how to port hardware developed under DOS over to Linux. Linux is a Unix-like multiuser, multitasking operating system for PC platforms with a 80386 or higher processor. Linux has made great inroads into the Unix market. The reason for this success is because Linux is free for all practical purposes and runs on the popular PC machines.

Linux began as a hobby for Linus Torvalds, a graduate student at Helsinki University, and it quickly caught the imagination of scores of kernel hackers all over the world. Today, you can obtain Linux free over the Internet or pay about US$50 for a Linux CD-ROM distribution kit. There are score of books explaining the features of Linux and a monthly Linux periodical as well. All the arguments for the choice of Unix as an operating system hold true for Linux. For the experimenter wanting to connect computers to external hardware for data logging and/or control, connecting to a machine running Linux (or Unix) is a nightmare; at least this is very true for those who have grown up with BBC micros, ZXs, and IBM PCs (running DOS). These machines allow you to connect to the required hardware with a simple BASIC program. However, only a single task at a time could run on these machines and any programming error could crash the system. These machines allowed the user complete control of the computer hardware without any supervisory control. With machines running Linux, this is no longer true. Errors in user programs under Linux cannot crash the entire machine. However, Linux requires special device drivers to connect any device to the computer. In order to understand how to interface Linux machines to external hardware, you must start with an understanding of the philosophy of device interfacing under Unix. Linux, like Unix, interfaces to hardware devices through a special set of files that are part of the file system. These files are called device files and are of two types: character devices files or block device files.

The use of a file-like interface to hardware devices promotes a uniform approach to data transfer. This also compels a device driver to be written in a uniform manner, presenting a uniform set of interfaces via the fops (file operations) structure. All hardware device-specific functions are encapsulated in the `open`, `close`, `read`, `write` amd `iocntl` modules of the fops structure. This enables you to use the same file-level access control mechanisms to control access to the hardware device. The next question is whether the I/O ports of the PC should be accessed directly. The answer is yes, but you must invoke system root privileges to directly access the hardware.

In this chapter, I will describe such a device driver, which was created for a 12-bit resolution, multichannel ADC system. In the second part of the chapter, I will show an application of this system: a weather data recording and display facility for the Internet.

A General-Purpose Data Acquisition System for Linux

This section describes a general-purpose data acquisition system for Linux. The end result is a simple hardware unit that, together with the `SanSon1gp_das` device driver described later in this chapter, can be hooked to a PC parallel port to record eight channels of analog voltage. The input range for each channel is 0–4.095V. The rate at which the data can be recorded is limited by the speed of the PC (for a 486 DX2 at 66MHz, the speed is about 160μs/sample, as compared to 220μs on a 386 at 20Mhz). The hardware used in this example is the MAX186 ADC interface for the parallel adapter, described in Chapter 8.

The SanSon device driver was developed in collaboration with my friend, Sunu Engineer. The acronym SanSon is a deep secret known only to our wives.

Testing Hardware on Linux

If adding hardware to a Linux machine requires you to add a specialized device driver, how can you run preliminary tests without the aid of the device driver? The answer is that, if you login with root privileges, all ports are accessible without any supervisory control. This allows the user to test the hardware, but puts the machine at the risk of a system crash.

The code in Listing 14.1 (`testport.c`) shows how to access I/O ports with root privileges. The parallel adapter DATA port address was determined by putting the machine in DOS mode — later in this chapter I will show how to obtain the parallel adapter DATA port address from within Linux. The function `ioperm()` is an important function. With this function, the program obtains permission to access requested ports. You can also use the similar function iopl(), but iopl() gives unlimited access to the program and should be avoided. Another feature that is important to note is the syntax of the macro outb(). DOS programmers are used to the similar function outportb(port_address, port_value), but with the gcc compiler, the structure is outb(port_value, port_address). The program in the following listing sends a

Listing 14.1   Access I/O ports with root privileges.

```
/*----------------------------------------------*/
/* testport.c */
/* compile as: gcc -O testport.c */
/* Execution is possible only as a superuser*/
/*----------------------------------------------*/

#include <asm/io.h>
#include <asm/segment.h>
#include <asm/system.h>
#include <unistd.h>

int ioperm();

#define port_add 0x378

void main()
{
    unsigned char test_value;
    int ret_value;

    ret_value=ioperm(data_port, 3, 1);

    if ( ret_value == -1 )
                    {
                    printf("Cannot get I/O permission\n");
                    exit(-1);
                    }
    outb(0x55,port_add);

    test_value=inb(port_add);
    printf("\nValue at Data port= %x\n", test_value);

    test_value=inb(port_add+1);
    printf("\nValue at the Status Port= %x\n", test_value);

    test_value=inb(port_add+2);
    printf("\nValue at Control Port= %x\n", test_value);
}
```

byte to the DATA port and reads it back. It also reads the contents of the STATUS and CONTROL registers. The value sent to the DATA port and the value read back should be the same, unless the port bits are being pulled externally.

The SanSon Device Driver

Now that you know how to access the parallel adapter port with superuser privileges, it is time to discuss the device driver.

As mentioned previously, it is important to have a device driver for the ADC under Linux so that the computer can operate the device efficiently (while in user mode) in a multitasking environment. It is not possible to detail the construction of a character device driver under Linux here, so I will mention only the most relevant features and the procedure we used when we wrote the SanSon device driver.

1. Write and test the code to evaluate the ADC performance with superuser privileges.
2. Write a character device driver for the ADC and link it to the kernel at compile time.
3. Write a set of functions that can be used at the user level to open the device and read the data in real time.

The following sections detail the implementation of this process.

Initial Steps

As mentioned in the section "Testing Hardware on Linux," we wrote some test code as superuser to access the parallel port directly using the `ioperm()` or `iopl()` calls in Linux, which allow the program to access a window of specified width into the I/O space. The code tests for the presence of a parallel port, identifies which port is to be used, tests for the presence of the device (i.e., the ADC), and reads data from the device directly using the Linux `inb()` and `outb()` calls.

As mentioned before, this testing can cause the system to hang (with a reboot as the only option), so you must be careful while doing this on a fully operational multiuser system. Also, it is better to use the `ioperm()` call rather than `iopl()`, because `ioperm()` allows only restricted access to I/O ports.

Once satisfied that the ADC is working as expected, move on to the process of constructing the device driver and integrating it into the kernel.

Structure of the Device Driver

The driver code is structured as follows:

- includes
- global variables

- adc_open: opens the device for reading if it is not already open (i.e., checks whether the variable adc_busy is set, and if not, permits the user program to open the device file corresponding to the ADC and read from it
- adc_close: closes the device and frees it
- adc_read: transmits the control byte corresponding to the minor device selected (Channel Chosen), reads an int from the ADC (the result of the conversion process, properly offset) and transfers it to the user space from the kernel space
- test_parallel_port: tests for the presence of a parallel port and returns the port address (Not Robust or Complete in the current implementation)
- test_adc: tests whether the ADC is connected and powered on in the specified parallel port
- adc_fops: declares and initializes structure
- adc_init: described later in this section

For the newer Linux kernels (1.2 onward), there is a directory called /usr/src/linux/drivers/char where the character device driver codes are placed. The driver has a drivername_init function, which, for the newer kernels, returns an int and takes a void, which serves as the point of linking to the kernel. In the source code mem.c (available on the companion code disk) directory /usr/src/linux/drivers/char, there is the function chr_dev_init in which one simply adds a devicename_init call and recompiles the kernel.

Listing 14.2 (ADC.c) is the complete device driver code, which contains the adc_init code. Listing 14.3 is the header file ADC.h used by ADC.c.

Linking the Device Driver into the Kernel

The process of recompilation of the kernel involves:

- cd to /usr/src/linux/drivers/char
- edit Config.in to include

  bool 'SanSon General Purpose ADC Support' CONFIG_SANSON_ADC
- Edit the Makefile to include

  ```
  ifdef CONFIG_SANSON_ADC
  L_OBJS += ADC.o
  endif
  ```
- cd to /usr/src/linux
- type make config and answer all queries appropriately
- type make clean; make dep
- type make zImage to be safe, and wait for the make to finish (this can take 45 minutes or more)

Listing 14.2   Device driver code.

```
/* ADC.c */

#include<linux/errno.h>
#include<linux/fs.h>
#include<linux/major.h>
#include<linux/kernel.h>
#include<linux/signal.h>
#include<linux/module.h>
#include<linux/sched.h>
#include<asm/io.h>
#include<asm/segment.h>
#include<asm/system.h>

#include "ADC.h"

extern int printk(const char* fmt, ...);
int adc_busy=0;
                        /*  continued  */
```

Listing 14.3   Header file used by ADC.c.

```
/* ADC.h */
#include<sys/ioctl.h>
#define ADC_MAJOR 31
#define ADC_MINOR 8

#define CHANNEL_0   0
#define CHANNEL_1   1
#define CHANNEL_2   2
#define CHANNEL_3   3
#define CHANNEL_4   4
#define CHANNEL_5   5
#define CHANNEL_6   6
#define CHANNEL_7   7

#define PARALLEL_1 0x378
#define PARALLEL_2 0x3bc
#define PARALLEL_3 0x278

#define ADC_BUSY 1
#define ADC_FREE 0

#define ADC_AVAILABLE 1
#define ADC_NOT_AVAILABLE 0
```

Listing 14.2 (continued)

```
unsigned short data_port,control_port,status_port;
static int adc_open(struct inode *  inode,struct file * file)
{
   unsigned int minor=MINOR(inode->i_rdev);

   if (minor>7) return -ENODEV;
   if (adc_busy==ADC_BUSY) return -EBUSY;

   adc_busy=ADC_BUSY;
   return 0;
}

static void adc_close(struct inode * inode,struct file * file)
{
   adc_busy=ADC_FREE;
}
static int adc_read(struct inode * inode,struct file * file,
                   char * buf,int count)
{
   unsigned char data[16];
   int adc_val;
   unsigned int temp_val;
   int temp1,temp2,temp3;
   unsigned int minor=MINOR(inode->i_rdev);

   if (count < 1) return -EINVAL;
   switch (minor)
   {
   case CHANNEL_0:
    temp1=0x8f;
    break;
   case CHANNEL_1:
    temp1=0xcf;
    break;
   case CHANNEL_2:
    temp1=0x9f;
    break;
   case CHANNEL_3:
    temp1=0xdf;
    break;
   case CHANNEL_4:
    temp1=0xaf;
    break;
   case CHANNEL_5:
    temp1=0xef;
    break;
```

- for safety, make a bootdisk with your old kernel
- copy the `zImage` file in directory `/usr/src/linux/arch/i386/boot` to `/vmlinuz` or whatever your kernel image is
- type `lilo` if you use LILO
- reboot
- if the kernel hangs, boot with the bootdisk, fix bugs, and go through the same steps again
- skip the `make clean` (after you have done it once) to speed up the compilation and linking process

---

Listing 14.2 (continued)

```
case CHANNEL_6:
 temp1=0xbf;
 break;
case CHANNEL_7:
 temp1=0xff;
 break;
default:
 temp1=0x8f;
 break;
}
cli();
for(temp2=0; temp2<8; temp2++)
{
   temp3= (temp1 << temp2) & 0x80;
   outb(temp3,data_port);
   temp3=temp3 | 1;
   outb(temp3,data_port);
   outb(temp3,data_port); /* this is to make the clk 50% duty cycle*/

   /* Duty cycle as measured with a 66 MHz 486 is 48% */
   temp3=temp3 & 0xfe;
   outb(temp3,data_port);
}
temp3=temp3 & 0x7f;
outb(temp3,data_port);

for(temp2=0; temp2<16; temp2++)
{
   temp3= 01;
   outb(temp3,data_port);
   data[temp2]=inb(data_port+1)&0x80;
```

- make the device files in /dev directory:

```
mknod /dev/adc0 c 31 0
mknod /dev/adc1 c 31 1
mknod /dev/adc2 c 31 2
mknod /dev/adc3 c 31 3
mknod /dev/adc4 c 31 4
mknod /dev/adc5 c 31 5
mknod /dev/adc6 c 31 6
mknod /dev/adc7 c 31 7
```

Listing 14.2 *(continued)*

```
      temp3=temp3 & 0xfe;
      outb(temp3,data_port);
      outb(temp3,data_port);
   }
   sti();
   adc_val=0;

   for(temp2=0; temp2<16; temp2++)
   {
      temp_val=( (unsigned int) data[temp2] & 0x00ff) << 8;
      adc_val= adc_val | ( (temp_val ^ 0x8000) >> temp2);
   }

   adc_val=adc_val>> 3;
   put_fs_word(adc_val,buf);
   /* printk("ADC: Input value from port: %d\n",adc_val); */
   return 1;
}

static struct file_operations adc_fops=
   {NULL,
    adc_read,
    NULL,
    NULL,
    NULL,
    NULL,
    NULL,
    adc_open,
    adc_close,
    NULL
   };
```

Listing 14.2 *(continued)*

```
int init_module(void)
{
   unsigned char rct_val,test_val-0x00;

   outb(test_val,PARALLEL_1);
   ret_val=inb(PARALLEL_1);
   if (ret_val==test_val)
   {
      data_port=PARALLEL_1;
   }
   else
   {
      outb(test_val,PARALLEL_2);
      ret_val=inb(PARALLEL_2);
      if (ret_val==test_val)
      {
         data_port=PARALLEL_2;
      }
      else
      {
         data_port=PARALLEL_3;
      }
   }
   printk("ADC: init module \n");
   if (register_chrdev(ADC_MAJOR,"adc",&adc_fops))
   {
      printk("Register_chrdev failed: Quitting\n");
      return -EIO;
   }
   else
   {
      printk("ADC: Device Registered\n");
      return 0;
   }
}

void cleanup_module(void)
{
   int busy=0;
   printk("ADC: Cleanup Module \n");
   if (adc_busy==ADC_BUSY) busy=1;
   if (busy) printk("ADC: Device busy, remove later\n");
   if (unregister_chrdev(ADC_MAJOR,"adc")!=0)
     {
         printk("ADC: Clean up module failed\n");
     }
```

After compilation and linking, the resulting kernel is installed in the system. The ADC device is connected to the parallel port and powered on. While booting, the kernel calls the adc_init function, which

- tests the parallel port to find the port base address,

- tests whether ADC is connected to the port and powered on, and

- if yes, registers the device with the kernel with a chosen major number (in this case 31).

Listing 14.2 (continued)

```
    else
    {
       printk("Clean up module succeeded\n");
    }
}
unsigned short test_parallel(void)
{
   unsigned char ret_val,test_val=0x55;
   unsigned short dport;

   outb(test_val,PARALLEL_3);
   ret_val=inb(PARALLEL_3);
   if (ret_val==test_val)
   {
      dport=PARALLEL_3;
   }
   else
   {
      outb(test_val,PARALLEL_1);
      ret_val=inb(PARALLEL_1);
      if (ret_val==test_val)
      {
         dport=PARALLEL_1;
      }
      else
      {
         outb(test_val,PARALLEL_2);
         ret_val=inb(PARALLEL_2);
```

Dynamically Installing Modules

Another way to include the device driver into the kernel is to dynamically insert it into a running kernel. This requires that the device driver contain two functions: `init_module` and `cleanup_module`. Dynamic installation into a running kernel is done through the `insmod ADC.o` command. The dynamic installation can be removed by the command `rmmod ADC`. This also requires that the string `kernel_version` be defined in the driver as:

Listing 14.2 (continued)

```
            if (ret_val==test_val)
            {
               dport=PARALLEL_1;
            }
            else
            {
               printk("No Parallel Port Available\n");
               return -EIO;
            }
         }
      }
   return(dport);
}

int test_adc(void)
{
   unsigned char cbyte;
   cbyte=inb(data_port+2);
   cbyte=cbyte&0xfe;
   outb(cbyte,data_port+2);
   cbyte=inb(data_port+1);
   cbyte=cbyte & 0x80;
   if (cbyte) return ADC_NOT_AVAILABLE;
   cbyte=inb(data_port+2);
   cbyte=cbyte|0x01;
   outb(cbyte,data_port+2);
   cbyte=inb(data_port+1);
   cbyte=cbyte&0x80;
   if (!cbyte) return ADC_NOT_AVAILABLE;
   return ADC_AVAILABLE;
}

int adc_init(void)
{
   int adc_stat;
```

```
static char * kernel'version=UTS'RELEASE;
```

If the kernel complains of a mismatched version, the module can be installed using insmod -f ADC.o. The init_module function does the job of testing and registering the ADC driver with the kernel.

Listing 14.2 (continued)

```
printk("\n\n");
printk("    General Purpose DAS : Gadre and  Engineer \n");
printk("    Copyright 1996  The Peshwe at IUCAA , Pune\n");

data_port=test_parallel();

if (data_port!=0x378 &&  data_port!=0x278 && data_port!=0x3bc)
{
   printk("Parallel Port Not Available\n");
   return -EINVAL;
}

printk("The Code detected %x as the parallel port in this machine\n",
       data_port);
adc_stat=test_adc();

if (adc_stat==1)
{
   printk("ADC: registering driver\n");

   if (register_chrdev(ADC_MAJOR,"adc",&adc_fops))
   {
      printk("Register_chrdev failed: Quitting\n");
      return -EIO;
   }
   else
   {
      printk("ADC: Device Registered\n");
      printk("  \t\t SanSon DAS  testing Successful \t\t \n\n");
      return 0;
   }

}

printk("ADC not connected\n");
printk(" \n\n");
return -EIO;
}
```

Using the ADC

Once the registration is complete, it is very simple to use the ADC. If you wish to use the user functions provided in file uadc.c (available on the companion code disk), simply use the call:

adc(CHANNEL_NUMBER, DEVICE_OPEN)

followed by any number of adc(CHANNEL_NUMBER, DEVICE_READ) calls and finally an adc(CHANNEL_NUMBER, DEVICE_CLOSE) call. Then repeat the process with any channel (0–7). The code as currently implemented allows only one user process to access any one channel of the ADC at a given time. The various channels on the ADC are configured as single-ended, unipolar inputs. The low-level way to access a channel is to open (with a system open call) the device file in the /dev directory corresponding to the required channel number. The files are named /dev/adc0-/dev/adc7 and are made using the mknod command as superuser:

mknod /dev/adc0 c MAJOR MINOR

Eight such files are made with the same major number (31 in this case) and different minor numbers (0–7) for the eight channels of the ADC. If the open returns true, one can read two unsigned chars from the buffer obtained using a read call on the file

---

Listing 14.4 Access a device using a user function call.

```
/* use_adc.c */
/* compile as : gcc -O use_adc.c */
/* Ordinary user mode */
#include "uadc.h"

int main(void)
{
   int ret_val;
   long i;

   ret_val=adc(CHANNEL_3,DEVICE_OPEN);
   if (ret_val==0) printf("Device Successfully Opened\n");
   for (i=0;i<=1000000L;i++)
   {
      ret_val=adc(CHANNEL_3,DEVICE_READ);
      printf("Data from device is %d\n",ret_val);
   }
   ret_val=adc(CHANNEL_3,DEVICE_CLOSE);
   return 0;
}
```

descriptor returned by an `open`. These two can be reassembled to form the actual output, after typecasting to an `int`. The device file must be finally closed with a `close(file descriptor)` call.

Listings 14.4 and 14.5 are sample programs using the user functions and accessing the driver at a system function level. Listing 14.4 (`use_adc.c`) shows how to access the device using a user function call. Listing 14.5 shows how to access the device using the system `open` call.

Using the DAS to Log Temperature

The data acquisition system described in this section records the ambient temperature of the surroundings using the circuit schematic below and the user code in Listing 14.6 (`log_temp.c`). The code accesses ADC channel 3 using the function `adc()`.

```
                        o +5V
                        |
                        /
                        \ R1 (2Kohm)
                        /
    Channel 3 input <------o
                        |
                       ---
                        ^
                       / \ LM336 Temp. sensor
                       ---
                        |
    ADC input GND    <-----o
```

```
/* log'temp.c */
/* uadc.c */
/* uadc.h */
```

Hosting a Weather Station on the WWW

In view of the increasing importance of the Internet as an information transfer medium, you may someday wish to implement a scheme that acquires data in real time and places it on the net. In this section, I will describe an extremely simple scheme to acquire experimental data using suitable hardware in a networked environment and distribute this data over a network. In the absence of any standards for dynamic information distribution, I have chosen to implement a design that is most importantly:

- simple — the required hardware and software is easily constructed and maintained, inexpensive, and readily available, and

- stable — the design must have a high degree of reliability.

The system described here relies on a compact data acquisition card that we have built, which plugs into a free parallel port on a PC together with a corresponding Linux

---

Listing 14.5 Access a device using the system open call.

```
/* sys_adc.c */
/* compile as : gcc -0 sys_adc.c */
/* used in Ordinary user mode */

#include <stdio.h>
#include <fcntl.h>
#include <sys/ioctl.h>
#include "adc.h"

int main(void)
{
    int fd;
    unsigned char  buf[5];
    int count=2;
    int i;

    fd=open("/dev/adc3",O_RDONLY);
    if (fd<0)
    {
        printf("Could not open device\n");
        abort();
    }
    else
    {
        printf("ADC Channel 3 opened\n");
    }
    for (i=0;i<=100;i++)
    {
        read(fd,buf,count);
        printf("Reading No:%d  %d \n",i,((int)buf[1]*256+(int)buf[0]));
        sleep(1);
    }
    close(fd);
    return 0;
}
```

driver described in the last section. The data is acquired through user-mode interfaces and converted to HTML format for distribution on the net by a daemon program.

```
/* Open adc device */
if(adc(channel,DEVICE_OPEN) != -4)
/* channel = ADC channel = 0 to 7 */
  {
  /* Read a data value from adc */
  data_point=adc(channel,DEVICE_READ);
```

Listing 14.6   Access ADC channel 3 using adc().

```
/* log_temp.c */
/* compile uadc.c with: gcc -c uadc.c */
/* then compile with: gcc -0 log_temp.c uadc.o */
/* used in Ordinary user mode */

#include <stdio.h>
#include <stdlib.h>
#include "uadc.h"

int main(int argc, char * argv[]))
{
  int data_point;
  float temp;
  int i;

  if (argc < 3)
  {
     fprintf(stderr,"Usage: argv[0] Number_of_minutes output_file\n");
     return -1;
  }

  if(adc(CHANNEL_3,DEVICE_OPEN))
  {
     outfile=fopen(argv[2],"w");
     for (i=0;i<atoi(argv[1]);i++)
     {
        data_point=adc(CHANNEL_3,DEVICE_READ);
        temp=(float)(data_point-2730)/10.0;
        fprintf(outfile,"%d  %f \n",i,temp);
        sleep(60);
     }
     adc(CHANNEL_3,DEVICE_CLOSE);
  }
  return 0;
}
```

```
    /* Close the adc device */
    adc(channel,DEVICE_CLOSE);
    }
else
    {
    /* If device could not be opened log error and exit */
    fprintf(logfile,"Could not open device\n");
    fclose(logfile);
    fclose(outfile);
    exit(-1);
}
```

Weather Information on the Web

Even small weather stations are extremely expensive devices. Many web sites use off-the-shelf devices to maintain local weather information on their web pages. Temperature, wind speed and direction, rain, and humidity are important weather parameters. Each of these parameters may need to be recorded at different rates.

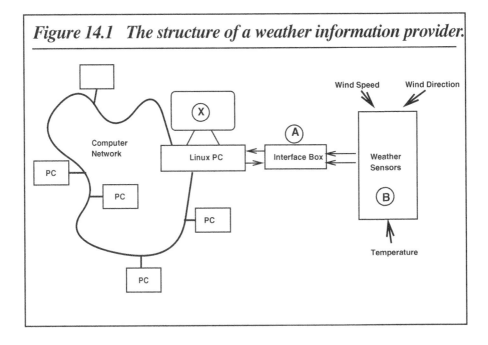

Figure 14.1   The structure of a weather information provider.

Listing 14.7   Record temperatures at one-minute intervals.

```c
/*adc_read.c*/
/*For weather data logging system*/

#include <stdio.h>
#include <stdlib.h>
#include <unistd.h>
#include <time.h>

/* Personal includes and defines */
#include "uadc.h"
#define USAGE "sampling_rate(in seconds)(0-300) num_samples(1-20)
               channel(0-7)"
/* End Personal includes */
/* Prototypes */
int adc(int,int);
/* End Prototypes */

int main(int argc,char *argv[])
{
    /* Sampling rate (in seconds) passed to sleep system call.
       and Number of samples */
    int sampling_rate,num_samples;

    /* Channel number of adc read from */
    int channel;

    /* Number of points in data file */
    int num_point=1;

    /* The integer value returned by adc */
    int data_point=0;

    /* The temperature obtained  by rescaling the integer value returned by
       adc */
    float temp;

    /* An index variable */
    int i;

    /* File pointers for input , output and log files */
    FILE * infile;
    FILE * outfile;
    FILE * logfile;

    /* POSIX type time_t for time system call */
    time_t init_time,current_time;
```

How Do You Do It?

The sketch in Figure 14.1 shows the plan. The PC marked X acts as the weather information server. This machine reads and logs weather data from the sensors (box B)
through the interface box (box A) at regular intervals. The sensors could be for
recording temperature, wind speed, wind direction, humidity, and rain.

Weather Data Logger Program

The prototype system collects weather data only for ambient temperature. The
weather data is recorded at intervals of one minute. The data is logged in a file with a
timestamp. At any instant, data corresponding to the last 15 minutes is maintained. A
daemon running at a frequency of once every 15 minutes does the job of processing
the data to produce a file in HTML format containing statistics, such as mean, maximum, and minimum for the period in question, along with diurnal data, such as the
maximum and minimum with timestamps.

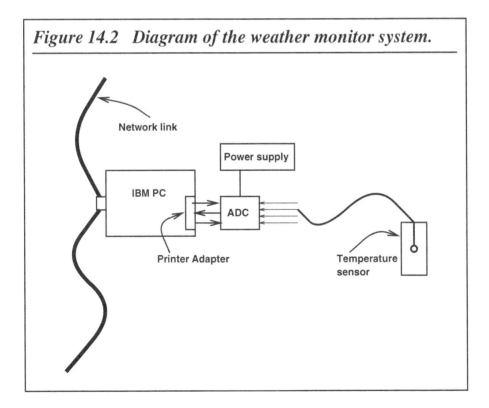

Figure 14.2   Diagram of the weather monitor system.

Listing 14.7 is a C program that records temperatures at one-minute intervals in a temperature data file. The file is maintained as a FIFO with a record of the last 15 readings. The weather monitor reads temperature data from an IC-based temperature sensor (Figure 14.2). The temperature sensor is connected to one of the channels of a multichannel ADC through the parallel adapter. The SanSon device driver is used by a user program running as a daemon on the Linux machine to record temperature readings every minute.

Listing 14.7 (continued)

```c
    /* A string for ctime system call to convert time_t type to a friendlier
       format */
    char *timestr;

    /* A string to read into and a char * to check the status of fgets */
    char instring[80];
    char *fget_stat;

    /* Parse and assign command line arguments */
    if (argc != 4 )
    {
        fprintf(stderr,"%s %s\n",argv[0],USAGE);exit(-2);
    }
    sampling_rate=atoi(argv[1]);
    num_samples=atoi(argv[2]);
    channel=atoi(argv[3]);
    /* Debugging purposes only */
#ifdef DEB
    fprintf(stderr,"Channel Number:%d\t Sampling Rate:%d Seconds \t
                Number of Samples : %d \n",
        channel,sampling_rate,num_samples);
#endif
    /* Get Starting time (Returns time as seconds elapsed since 00:00:00
       January 1,1970 */
    init_time=time(NULL);

    /* ctime converts time to a string of form Sun Jun 28 00:00:00 1996  */
    timestr=ctime((const time_t *)&init_time);

    /* Write starting time into logfile */
    logfile=fopen("adcerr.log","w");
    fprintf(logfile,"Start Time: %s",timestr);
    fflush(logfile);

    /* Open the output file (temperature.dat) */
    outfile=fopen("temperature.dat","w");
```

Listing 14.8 is a PERL script that processes the information available in the temperature data file to produce an output file in HTML format containing the current average temperature and the minimum and maximum temperature records for the day. Note that the PERL script filename on the DOS companion disk has been truncated to `read_dat.per`.

Listing 14.7 (continued)

```
/* Start The infinite loop */
for(;;)
{
    /* Open adc device */
    if(adc(channel,DEVICE_OPEN) != -4)
    {
        /* Read a data value from adc */
        data_point=adc(channel,DEVICE_READ);

        /* Set  time of data read */
        current_time=time(NULL);

        /* Convert to an  easier format */
        timestr=ctime((const time_t *)&current_time);

        /* Rescale the data as per sensor requirements */
        temp=(float)(data_point-2730)/10.0;

         /* Debugging purposes only */
#ifdef DEB
 fprintf(stderr,"Temperature:%3.1f \t Time:%s\n ",temp,timestr);
#endif
        /* Close the adc device */
        adc(channel,DEVICE_CLOSE);
    }
    else
    {
        /* If device could not be opened log error and exit */
        fprintf(logfile,"Could not open device now \n");
        fclose(logfile);
        fclose(outfile);
        exit(-1);
    }

    /*  Initially at startup write num_samples data points (Sampled at a
        sampling_rate(in seconds) interval ) to the output file . This is
        read by another program which averages, finds minimum and maximum,
        and outputs a HTML file containing the relevant data */
```

Listing 14.9 is a shell script to initiate the data acquisition. Note that the shell script filename on the DOS companion disk has been truncated to `start_lo.sh`.

Additional Features

Apart from adding more sensors, the system would be more useful if you added longer statistics about the weather parameters (e.g., the temperature readings for a month or more or daily maximum and minimum temperatures for a month). You could easily make such modifications to the present system.

Listing 14.7 (continued)

```
if (num_point<=num_samples)
{
    /*  Just a precaution in case outfile is open to flush it */
    fclose(outfile);

    /* Open output file in append mode */
    outfile=fopen("temperature.dat","a");

    /* Output the temperature as a float and time of reading as a string */
    fprintf(outfile,"%3.1f %s",temp,timestr);
    fclose(outfile);

    /* Increment number of points written to file */
    num_point++;
}
else
{
    /* If the number of points in file is greater than num_samples */
    if (num_point>num_samples)
    {
        /* Open the output file (temperature.dat) as read only
        and a temporary file (temperature.datt) . Copy the last fourteen
        lines  of temperature.dat to the temporary file */

        infile=fopen("temperature.dat","r");
        outfile=fopen("temperature.datt","w");
        for(i=0;i<=14;i++)
        {
            fgets(instring,32,infile);
            if (i>0) fprintf(outfile,"%s",instring);
        }
        fclose(infile);
        fclose(outfile);
```

Listing 14.7 (continued)

```c
        /* Delete original file (temperature.dat) and copy the temporary
           file to temperature.dat */
        infile=fopen("temperature.datt","r");
        outfile=fopen("temperature.dat","w");
        while ((fget_stat=fgets(instring,32,infile))!=NULL)
        {
            fprintf(outfile,"%s",instring);
        }
        fclose(infile);

        /* Delete tempfile */
        unlink("temperature.datt");

        /* Write the current data point and time stamp to
           temperature.dat. This ensures that temperature.dat
           always contains the most recent fifteen temperature values */
        fprintf(outfile,"%3.1f %s",temp,timestr);

        /* Flush and close output stream */
        fflush(outfile);
        fclose(outfile);
    }   /*  End if (num_point>num_samples) */
  }    /* End if (num_point<=num_samples) */

  /* Sleep for sixty seconds ( Sampling at one minute intervals) */
  sleep(sampling_rate);
 } /* Infinite Loop */
}   /* Close Main */
```

Listing 14.8   Produce an HTML file using PERL.

```perl
#!/usr/bin/perl
#read_data_average.perl

# Number of Sample points to make running average of is supplied as a
# command line argumnt
$num_samples=$ARGV[0];

# Set minimum and maximum to arbitrarily high and low values respectively
$min_temp = 100.0;
$max_temp = -100.0;
@minimum=split(/ +/,localtime());
$min_temp_time=$minimum[3];
$min_temp_date=$minimum[1]." ".$minimum[2];
@maximum=split(/ +/,localtime());
$max_temp_time=$maximum[3];
$max_temp_date=$maximum[1]." ".$maximum[2];


# Infinite Loop to run as a daemon
for(;;)
{
   # Open temperature.dat for input (contains $num_samples lines of the
   # format temperature Sun Jul 28 00:00:00 1996

   open(INFILE,"<temperature.dat") || die "Can't open inputfile";

   # Read all the data into a single array
   # Each element of array contains one line of file

   @inputs=<INFILE>;
   close(INFILE);

   # $num_lines contains number of lines in the file
   $num_lines=@inputs;

   # Check to see if the number of lines is greater than number of sample
   # points to average over
   if ($num_lines >= $num_samples)
   {
      # If true then  average over the temperature field and also find the
      # minimum and maximum values for temperature.

      # Initialize counter,sum and average
      $counter = 1;
      $sum = 0.0;
      $average = 0.0;
```

Listing 14.8 (continued)

```
while($counter <= $num_samples)
{
   # Split  each element of input array on whitespace
   @data = split(/ +/,$inputs[$num_lines-$counter]);

   # Extract the first field as temperature
   $temperature = $data[0];
   $data_time = $data[4];
   $data_date= $data[2]." ".$data[3];

   if ($data_date ne $min_temp_date)
   {
      $min_temp_date=$data_date;
      $min_temp=$temperature;
      $min_temp_time=$data[4];
      $max_temp_date=$data_date;
      $max_temp=$temperature;
      $max_temp_time=$data[4];
   }
   # Find the minimum and maximum temperature of num_sample points
   if ($temperature < $min_temp)
   {  $min_temp = $temperature;
      $min_temp_time=$data_time;
      $min_temp_date=$data_date;
   }
   if ($temperature > $max_temp)
   {  $max_temp = $temperature;
      $max_temp_time=$data_time;
      $max_temp_date=$data_date;
   }
    $sum += $temperature;
   $counter++;
}

# Compute average
$average = $sum/($counter-1);
$current_time=localtime;

# Open output file in HTML format  (Weather.html)
open(OUTFILE,">Weather.html") || die "can't write output HTML file\n";
print OUTFILE "<HTML><TITLE>\n";
print OUTFILE " The Weather at IUCAA\n";
print OUTFILE "</TITLE>\n";
print OUTFILE "<BODY bgcolor=\"\#000000\" text=\"\#ffffff\
               "link=\"\#00ff00\" alink=\"\#ff0000\
               "vlink=\"\#ffff00\">\n";
```

Listing 14.8 (continued)

```
    print OUTFILE "<CENTER>\n<HR>\n";
    print OUTFILE "<H1>The Weather at IUCAA </H1>";
    print OUTFILE "<HR></CENTER>\n";
    print OUTFILE "<FONT text=\"\#ff0033\
                   " >Time of record (IST) is $current_time</FONT>\n<BR>\n";
    printf OUTFILE ("<FONT text=\"\#ff0033\
                    " >Average Temperature for the past $num_samples
                    minutes:\t %3.1f degrees Celsius</FONT>\n",$average);
    printf OUTFILE ("<FONT text=\"\#ff0033\
                    " ><HR>\nMinimumTemperature:%3.1f degrees Celsius at
                    $min_temp_time on $min_temp_date\n<HR>\nMaximum
                    Temperature:%3.1f degrees Celsius at $max_temp_time on
                    $max_temp_date<HR></FONT>\n",$min_temp,$max_temp);
    print OUTFILE "<BR>\n";
    print OUTFILE "<BR>\n";
    print OUTFILE "<BR>\n";
    print OUTFILE "<A HREF=\"http://iucaa.iucaa.ernet.in/~ilab/
                    instrumentation.html\">Back to Instrumentation</A>\n";
    print OUTFILE "</BODY></HTML>\n";
    close(OUTFILE);

    # Move Weather.html ~ilab/public_html
    system("mv  Weather.html ..");
  }
  else
  {
    sleep(15*60);
  }
  # End of If $num_lines >= $num_samples condition

  sleep(15*60);
}
# End infinite loop
```

Listing 14.9   Shell script to initiate data acquisition.

```
#!/bin/sh

ccho "Cleaning evrything\n"
make clean

echo "Starting Temperature reader  \n"
./adc_read 60 15 7 &
sleep 15m

echo "Starting HTML constructor \n"
./read_data_average.perl 15 &
```

Appendix A

PC Architecture

Introduction

The IBM PC was introduced in August 1981. At that time it was a unique, high-performance computer for the general public. It was also the first personal computer from IBM. The computer was designed such that additional hardware could be added if the user desired. The computer offered performance that, until then, was not available to the general public. IBM also allowed other manufacturers to duplicate (clone) the PC design. Soon a multitude of PC varieties was available on the market, some offering improved performance and, most significantly, some less expensive than the original IBM model.

The original PC used the 8088 processor from Intel. The 8088 processor is part of the x86 family of processors. In 1983, IBM introduced the PC/XT and in 1984, the PC/AT. The PC/XT (XT for extended technology) was essentially a PC with a hard disk. The PC/AT (AT for advanced technology) modified the PC architecture substantially. To begin with, it used the much more powerful 80286 processor. Also, the data bus inside the computer was increased from 8 to 16 bits. Computers must fetch code and data from memory, and a 16-bit bus enabled the processor to fetch the same data in half the time by doubling the width of the data bus. The PC/AT also added more hardware features, as detailed later in this appendix.

The PC/AT (or simply the AT) increased performance by a factor of 10. If the original PC took 10 seconds to run a program, the AT did it in one second. As more powerful processors in the x86 family were introduced by Intel, newer and more powerful PCs were designed and produced. However, even with the introduction and availability of higher performance PCs, the basic structure hasn't changed from the original PC. This is because, to keep newer machines compatible to the large base of DOS programs, manufacturers have been forced to retain operational similarity with the PC. To understand the operation of the parallel port, it is very important to understand the architecture of the PC and, more importantly, the architecture of microprocessor systems in general. In the following sections, I'll described the architecture of the IBM-compatible PC.

Understanding Microprocessor Systems

To appreciate the design of PC compatibles, it is useful to understand how a general-purpose computer circuit is structured. Every computer has five essential components:

1. Arithmetic Logic Unit (ALU);

2. Control unit (CU);

3. Memory;

4. Input device; and

5. Output device.

In most modern computers, the microprocessor combines the functions of the ALU and the CU. A microprocessor also needs a few registers and some control circuitry. A register is a latch that can be used to store information and read back information. A register, unlike a memory location, does not have an address — it is accessed by a symbolic name. The number of registers and the complexity of the ALU, CU, and control circuitry vary from microprocessor to microprocessor. Simple microprocessors may have only a few registers, and some advanced microprocessors may contain hundreds of registers.

By itself, a microprocessor cannot do anything useful. The designer must add memory to store user programs, user data, and program results and I/O devices to communicate with the external world (like a keyboard, video terminal, etc.). A microprocessor circuit works by fetching user programs from the external memory, decoding the program element (called the instruction), and executing the instruction.

To fetch an instruction from memory, the processor must have a way of signaling external memory and indicating whether it needs to read data or write data. The control bus provides a means of accessing external memory. The CPU (or the microprocessor) also needs an address bus that carries addressing information. A component called the data bus transfers data in and out of the CPU to registers, memory, or ports.

Figure A.1 shows the essential blocks of a simple microprocessor system. The microprocessor system works by reading the program stored in memory and executing the program instructions.

Every instruction is composed of an opcode and an optional operand. Simple instructions only have an opcode; more complex instructions have one or more operands or arguments. An example of an instruction with operands is the add instruction: add x, y. The instruction is composed of the opcode add and the operands or arguments on which the add operation is performed, x and y. After the two numbers are added, they must be stored somewhere. This instruction could transfer the result back into one of the variables, x. Thus, add x, y would amount to $x = x + y$.

The memory device is used not only to read program instructions, but to store the results of the program, variable values, etc. Eventually the program will generate data that must be sent out of the system through an output device, such as the screen, printer, or modem.

Memory is addressed though the address bus. Typically, microprocessors have between 16 and 32 address lines (some modern varieties have even more), resulting in an address space of from 65,536 to more than 4,294,967,000 locations. Each memory location can store data. The range of the data depends upon the width of the data bus. Typically, data buses come in widths from 8 to 32 or even 64 bits. The number of data bits is used as an indicator of the power of the microprocessor, which is often classified as an 8-, 16-, or 32-bit processor.

Figure A.1   *Block diagram of a simple microprocessor system.*

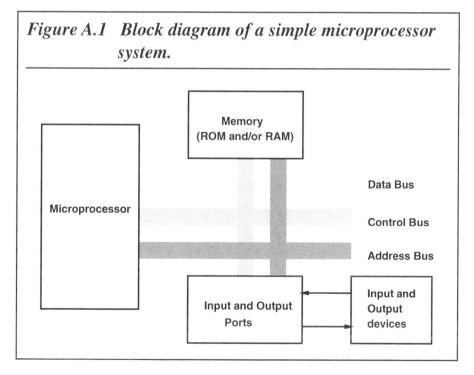

The control bus of a microprocessor has signals that regulate smooth data transfer in to and out of the processor to memory or to the ports. Because only one data bus connects to so many ports, devices, and memory locations, it is important that the data bus talks to only one component at any time; otherwise, conflicting voltages applied to the data bus from independent devices could result in a condition called bus contention, which could damage the processor or other components of the circuit.

A general-purpose microprocessor system has at least two types of memory: memory for permanent storage, called ROM or EPROM, and memory whose contents can be modified at any time, called Random Access Memory (RAM). ROM or EPROM is used to store the startup program, and RAM is used to load user programs from storage devices such as the disk. When you power-up the PC, the permanently stored program in the PC checks memory and other I/O devices and loads the operating system. Subsequently, when the user wants to execute a program of choice, the program is loaded from the disk into RAM and executed.

A microprocessor in an embedded system, such as the processor in a washing machine, doesn't have RAM to store user programs. The processor in such systems is expected to execute a program stored in the system ROM or EPROM that controls the device.

Bigger and bigger programs are being written, and demand for processors with large memory-handling capacity is increasing. When the PC was introduced in 1981, contemporary home computers were limited to 64Kb of memory, and a PC with 640Kb of memory was held in awe. It seemed that with 640Kb of memory, one couldn't ask for more. Today, it is impossible to think of running common application programs on this, and PCs with as much as 64Mb (a 100-fold increase) are now common.

Accessing Ports and Memory

Given a microprocessor that distinguishes between memory devices and I/O port devices, there must be a signal that indicates the nature of the device that the microprocessor wants to communicate with (Table A.1). Then there must be set of signals that indicate the nature of the activity (e.g., whether the microprocessor is reading the device or writing to it). The four signals shown below indicate the nature of the activity. The signals are shown with an asterisk to indicate that the active level of the signals is logic 0.

IOR* to indicate that the microprocessor will read from an input port whose address is set on the address bus;

Table A.1 Port addresses on the PC/AT.

Port address	Device
000h–00fh	First DMA chip (8237)
020h–021h	First interrupt controller (8259)
040h–043h	Interval timer (8253)
060h–063h	Keyboard controller (8042), 8255-PPIO on PC
070h–071h	Real-time clock
080h–083h	DMA page register
0a0h–0afh	Second interrupt controller
0c0h–0cfh	Second DMA chip
0e0h–0ffh	Reserved
100h–1ffh	Available for expansion
200h–20fh	Game adapter
210h–217h	Reserved
220h–26fh	Available for expansion
278h–27fh	Parallel port
2b0h–2dfh	EGA adapter
2f8h–2ffh	COM2
300h–31fh	Prototype adapter
320h–32fh	Available for expansion
378h–37fh	Parallel port
380h–38fh	SDLC adapter
3a0h–3afh	Reserved
3b0h–3bfh	Monochrome adapter
3c0h–3cfh	EGA
3d0h–3dfh	CGA
3e0h–3e7h	Reserved
3f0h–3f7h	Disk controller
3f8h–3ffh	COM1

IOW* to indicate that the microprocessor will write to an output port whose address is set on the address bus;

MR* to indicate that the microprocessor will read from a memory location whose address is on the address bus;

MW* to indicate that the microprocessor will write to a memory location whose address is on the address bus.

The microprocessor initiates the reading or writing activity and issues the relevant signals. Each activity is for a fixed amount of time that depends upon the microprocessor frequency of operation. Microprocessors operating at different frequencies issue signals of different duration.

Figure A.2 shows the timing diagram for a memory read operation. The CLOCK signal, the MR* signal, and the state of the ADDRESS and DATA buses are shown. This diagram shows the execution part of the instruction. The execution of this instruction would be preceded by the fetch and decode instructions, which I have not shown.

Figure A.2 shows that the memory read operation takes five clock cycles. During the first clock cycle, the microprocessor puts the source address of the memory loca-

Figure A.2   Timing diagram for a memory read operation in a hypothetical microprocessor system.

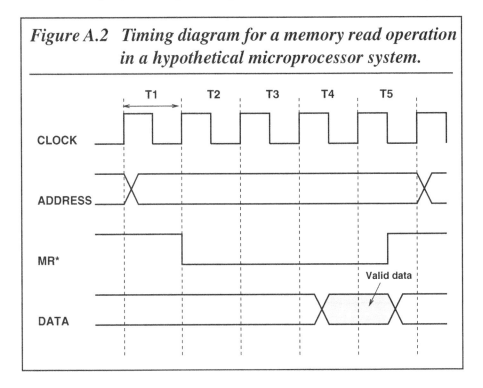

tion. During the second cycle, the MR* signal is asserted. It is important to note that the assertion levels of all the signals (MR*, MW*, IOR* and IOW*) are logic 0. The assertion of the MR* signals must be used to enable the memory location that would then drive the data bus. The memory device has the whole of T2, T3, and about half of T4 to put data on the data bus. Data from the memory device must be available after the rising edge of the T4 clock signal and must remain stable till the MR* signal is deasserted. The actual time for all the signals would be translated from the duration of the clock signal together with a table provided by the microprocessor manufacturer that lists the minimum and maximum worst-case timings. These values are used by the circuit designer to determine whether the selected memory device meets all the requirements and whether it will perform without failing in the system.

A similar timing diagram in Figure A.3 is shown for a port write operation. Here, the active signal is IOW*. The processor sets up the address of the destination port during T1. During T3, it sets up the data that is to be written into the port and asserts the IOW* signal. The data is held stable through T4 and a little time after that. The IOW* signal is deasserted at the end of T4. The rising edge of the IOW* signal would be used to latch the data into the destination port.

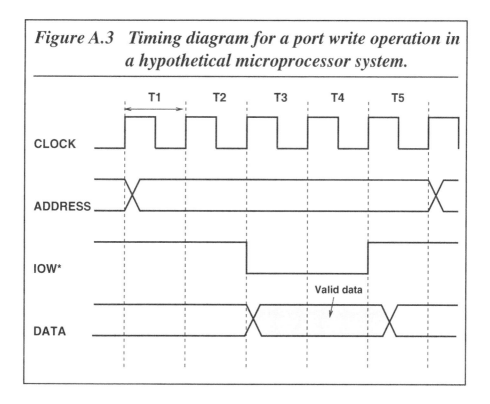

Figure A.3 **Timing diagram for a port write operation in a hypothetical microprocessor system.**

Support Sections of a PC

The computer system needs other peripheral devices to do the many things the micro-processor cannot do on its own. These support functions are performed by IC devices with some additional circuitry. Many of these ICs communicate with the CPU over the system bus as ports of the CPU. These chips are selected for optimum perfor-mance and programmability from among a wide range of components made by vari-ous manufacturers. Most of these support chips were originally designed by Intel. These chips and their functions are:

Peripheral Programmable I/O (PPIO) controller   This is the chip (8255 in the original PC and the AT) that the CPU uses to control other devices of the PC as well as to interface the PC keyboard. This chip has many input and output pins used to control the CPU speaker, read data generated by the keyboard, etc. In early PCs, the CPU used this chip to read system information at the time of the boot-up sequence. In modern PCs, this chip and many others have been replaced by a single, large multi-function IC.

Programmable Interrupt Controller (PIC)   The PIC is the heart of the PC inter-rupt structure. PIC is used by the CPU to respond to a large number of hardware inter-rupts generated by the other sections of the PC. Essentially, the PIC is a sort of interrupt multiplexer. The CPU of the PC has an interrupt line, but with the addition of PIC, which has many interrupt inputs (and one interrupt output connected to the CPU interrupt), the PC can respond to many more interrupts. The PIC is used by the serial ports (RS-232), floppy disk controllers, PC timer, and other components to grab the attention of the CPU. For every interrupt input of the PIC, the PC maintains a list of interrupt vectors. An interrupt vector is the address of a special subroutine called the Interrupt Subroutine. When the CPU receives an interrupt, it looks at the interrupt vector list and gets the address of the Interrupt SubRoutine (ISR) that needs to be exe-cuted in response. The original PC and the XT were equipped with a single PIC, which could handle up to eight interrupt inputs. In the AT and other machines, there are two PICs allowing up to 15 interrupt inputs.

The Direct Memory Access (DMA) controller   DMA is a concept that allows direct access of the memory by some devices without the involvement of the CPU. Data transfer using DMA is extremely fast. The DMA controller facilitates such transfers between the memory and the ports. DMA is used by floppy and hard disk controllers.

Programmable Interval Timer (PIT)   This device is used to generate timing sig-nals at regular intervals. These timing signals are used to update the current time of day. The PIT used in the PC has three channels. One of the other channels of the PIT is used to generate tones on the speaker under control of the PPIO chip.

Video controller   This section of the PC interfaces the monitor to the CPU and allows graphics and text output on the screen.

All of these PC components communicate with the CPU over the system bus.

PC System Bus Signals

All computer systems (general or embedded) need a system bus. The system bus is a name for the collection of signals that include the data bus, address bus, control bus, and other important signals of the computer system. In a general-purpose computer system such as the PC, the system bus is important because it allows the user to expand the operation of the system according to specific requirements and to tailor the system to specific needs. With the system bus, the user can add memory, add new devices, or even change the complete character of the system. For example, it is very common to take a PC, remove things that are not needed, add required ports or devices, and convert the system to a dedicated embedded system. Without the system bus, this type of conversion would be impossible.

When the PC was introduced, it had a system bus without a specific name. With time, new features were added and the system bus changed features as well as name. Today, many types of system busses are available on the PC motherboard. The most common ones are: ISA, EISA, and PCI. Some machines have more than one system bus, providing a combination of ISA, EISA, and PCI. In this section, I will confine the discussion to the good old nameless system bus that was a forerunner to the ISA bus.

The system bus is accessible through 62-pin card slots on the PC motherboard. These slots are used to insert circuit boards with a mating card edge connector. Different computers have different numbers of card slots. The original PC had five such identical slots. PCs with as few as three and as many as seven slots are available. Obviously, the PC with more slots has more options for expansion. The card slots have 62 signals, including signals for the data bus, address bus, control bus, and power supply bus. All the signals are TTL compatible and are generated by the microprocessor and other system components, such as the interrupt controller, DMA controller, etc. The signals and their functions are:

D0–D7 (I/O)   The data bus. The 8-bit data bus is bidirectional and is used for data transfer from and to the adapter cards that fit into the card slots.

A0–A19 (O)   The address bus has 20 bits and indicates the address of the data transfer between the CPU and other devices or the DMA controller and other devices.

IOW* (O)   This signal is generated either by the processor or the DMA controller to indicate data transfer to the addressed destination port is in progress.

IOR* (O)   This signal, generated by the processor or the DMA controller, indicates that data is read from the addressed port.

MEMW*(O)   This signal, generated by the CPU or the DMA controller, indicates that the CPU or the DMA controller wants to write data into the addressed memory location.

MEMR* (O)   This signal, generated by the CPU or the DMA controller, indicates that the CPU or the DMA controller wants to read data from the addressed memory location.

RESET DRV (O)   This signal provides the reset signal to ports and other devices during power up or during a hardware reset. It is an active high signal.

IRQ2–IRQ7 (I)   These are the interrupt inputs to the Programmable Interrupt Controller (PIC) chip on the motherboard.

CLK (O)   This is the highest frequency available on the card slot and is three times the OSC frequency.

OSC (O)   This is the clock signal to which all the IOW*, and other strobe signals are referenced to. It has a frequency between 4.77MHz on the original PC to 8MHz on newer PCs.

ALE (O)   The is the Address Latch Enable signal. During a transfer to or from the CPU, the CPU places the address on the address lines. The original CPU had the lower eight address lines multiplexed with the eight data bits. The ALE signal is a demultiplexer signal for the address information. On the system bus, the address and the data bits are already demultiplexed and the ALE signal is only used as a synchronization signal to indicate the beginning of a bus cycle.

TC (O)   This signal is generated by the system DMA controller to indicate that one of the channels has completed the programmed transfer cycles.

AEN (O)   The AEN signal is generated by the DMA controller to indicate that a DMA cycle is in progress. A DMA cycle could involve a port read and a memory write. However, the port address on the expansion card should not respond to the port read bus cycle if it is not intended. By using the AEN signal, the card circuit can detect whether the bus cycle is issued by the CPU or the DMA controller and respond accordingly. A high AEN indicates a bus cycle issued by a DMA controller.

I/O CH RDY (I)   This signal is used by the card circuit to indicate to the CPU or the DMA controller to insert wait states in the bus cycle. Up to 10 clock cycles can be inserted.

I/O CH CK* (I)   This signal can be used by the circuit on a plug-in card to indicate an error to the motherboard. An NMI hardware interrupt corresponding to INT2 is generated by the motherboard circuit in response to a low I/O CH CK* signal.

DRQ1–DRQ3 (I)   This is an input signal to the DMA controller on the motherboard. When a port device wants to transfer data to and from memory, it can use the DMA transfer cycle. The operation of the DMA transfer cycle is controller by the DMA controller. DRQ1 to DRQ3 are the three inputs to the DMA controller. At reset, the system ROM BIOS puts DRQ1 at the highest priority and DRQ3 at the lowest. The DMA controller has four channels, DRQ0 is used on the motherboard to generate dummy read cycles to refresh dynamic memory.

DACK0*–DACK3* (O)   These are the four status outputs of the DMA controller that indicate the acceptance of the DRQ request. The DMA transfer cycles begin after the DACK* line is put to 0.

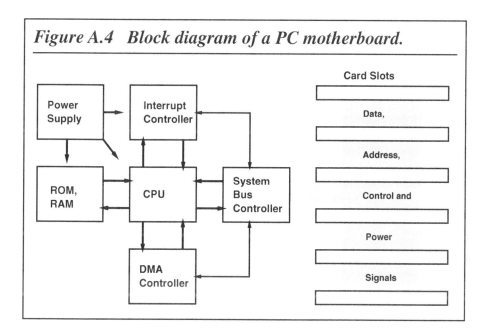

Figure A.4   Block diagram of a PC motherboard.

Power supply   The motherboard provides +5, +12, −5, and −12V voltages to the card slots. The +ve voltages are guaranteed to be within ±5% of their nominal values and the −ve voltages between ±10%.

The PC Ports

Figure A.4 shows the structure of the original PC motherboard. Note that the design does not include display, serial, or parallel adapters. This is because the original PC had all these interface devices in the form of plug-in cards that fit in the expansion card slots.

The expansion slots were connected to the motherboard through what is called the system bus controller. As PCs evolved, many of these devices were integrated on the motherboard. The system bus also was modified into a local bus and a system bus. More and more components were connected to the CPU through the local bus. The system bus remained to provide compatibility with the original PC and XT and also to provide system expansion capabilities to the user.

During the late 80s, the system bus was standardized as the Industry Standard Architecture (ISA) bus. A little later, a newer system bus, called the EISA bus, was offered for operation on the PC. A still newer bus called the Peripheral Component Interconnect (PCI) bus is now popular.

Almost all the peripheral devices, such as the display adapter, parallel port, serial port, and disk controller have been integrated on the motherboard, thereby significantly reducing the size of the PC. Figure A.5 shows the block diagram of a modern PC. The ports required by the devices are mapped onto the system bus.

Example of a Typical Interface Circuit

Consider a case where you needed to build a simple I/O expansion port (much like a parallel port) with eight bits of input and eight bits of output. You will map these ports in the 300h–31fh I/O address range with a facility to change the address with the help of DIP switches within the 300h–31fh range. The advantage of such a scheme is that the ports can be used even in a system with other cards using the same address range. If there is another card in the system with four ports addressed between 300h and 303h then, with the help of the DIP switches, you could relocate the port addresses on the card to be, say, 304h. Figure A.6 shows the block diagram of this simple I/O board.

Instead of using discrete I/O chips (like the ones used in the preceding example), designers often use programmable I/O chips. One of Intel's most popular I/O chips is the 8255-PPIO. The original PC used an 8255 chip on the motherboard to control the keyboard and speaker, read system DIP switches (used to encode system configuration information at boot time), etc. The 8255 was mapped at the 60h–63h I/O address in the original PC.

Figure A.7 shows the circuit schematic of the PC-based digital Input and Output card. J1 is the PC system bus interface. It contains the address, data, control, and power signals (15V and gnd). The data bits are numbered D7–D0, and the address bits are numbered A19–A0. The signal names are shown inside the J1 connector, and the numbers outside are the pin numbers. The connector has 2 sides, A and B. The address and data bits are on side A of the connector; the control signals are on side B. The circuit uses the following signals from the system bus:

D0–D7   The eight data bits.

A9–A2   Eight of the 20 address bits. For I/O interfacing, the processor uses only the lower 16 address lines A0–A15. In the PC design, of these 16 address lines, only the lower 10 address lines, A0–A9, are decoded. Thus, the number of possible ports in the PC is limited to 1,024 ports (1,024 input ports and 1,024 output ports). Of the 10 address lines in this example, I use only the higher eight address bits. Because the A0 and A1 address bits are not used, for any setting of the A9–A2 address bits on the

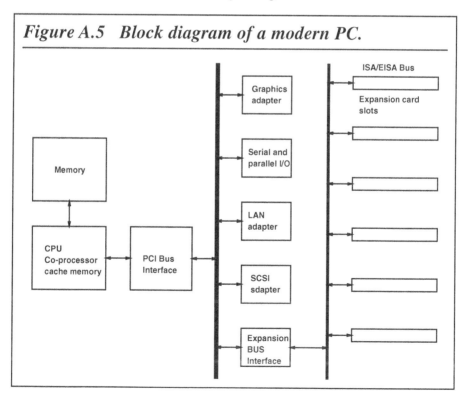

Figure A.5   Block diagram of a modern PC.

card, each port would have four possible addresses (e.g., if you set the address of the card to 300h, the ports would still respond to 301h, 302h, and 303h. (This scheme is not advisable in real situations — I have chosen it here for simplicity at the cost of wasted port addresses.)

AEN   This is the signal that indicates whether the current bus cycle is a DMA or a non-DMA cycle. The card needs to respond to only non-DMA bus cycles. I use the AEN signal to disable the card when the AEN signal is active.

IOW* and IOR*   These are the signals that indicate whether the bus cycle is to transfer data to or from the port.

+5V and gnd   These lines power the circuit.

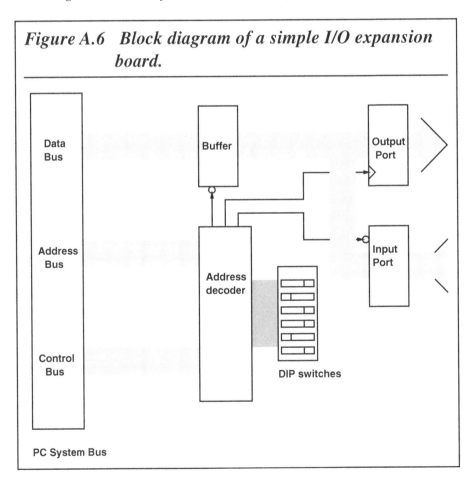

Figure A.6   Block diagram of a simple I/O expansion board.

Figure A.7 Circuit schematic of a simple I/O expansion board.

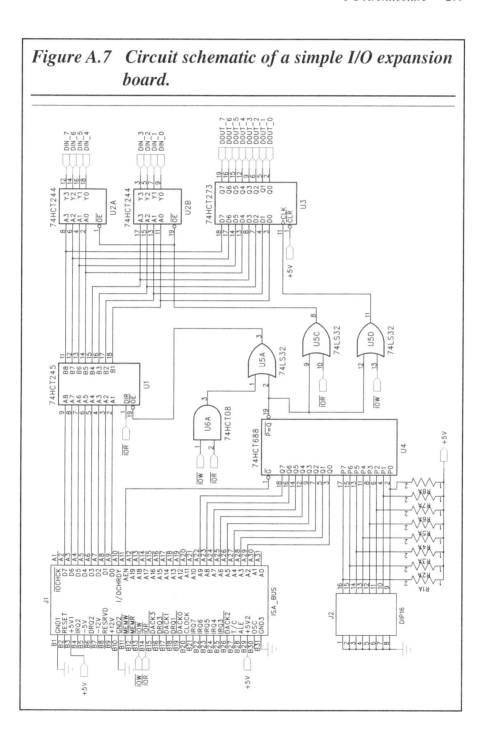

IC U1 (74HCT245), a bus transceiver, connects to the data bits of the system bus. This IC has the direction control bit DIR and the master enable bit OE*. When the DIR bit is low, the IC transfers data on the B side to the A side. When it is high, data on the A side is transferred to the B side. This IC is used to isolate the data bus from the rest of the circuit. With a buffer IC between the data bus and the rest of the I/O ports, the system bus is offered a constant load and is isolated from the rest of the circuit.

The address decoder circuit is composed of the IC U4 (74HCT688), an 8-bit comparator circuit. The IC has the master enable input pin $\overline{G}$ and active low output $\overline{P = Q}$. The comparison inputs are Q7–Q0 and P7–P0. When all the Q bits match with the corresponding P bits, the output goes low, otherwise the output is high. In this circuit, the P inputs are driven by a set of eight DIP switches and the Q inputs are connected to the A9–A2 address bits. The DIP switches are set to the required address combination. When the $\overline{G}$ input connected to the AEN signal of the system bus is low, and address bits A9–A2 match the DIP switch setting, $\overline{P = Q}$ goes low. The output of the comparator IC is further qualified with the IOW* and IOR* signals. When either of these signals is low, which indicates that a port I/O bus cycle is in progress, the output of gate U6A goes low. When U6A output is low and $\overline{P = Q}$ is also low, the output of gate U5A goes low. This indicates that an I/O cycle for the circuit is in progress. This signal is used to enable the U1 data bus buffer. Depending upon which signal, IOR* or IOW*, is low, either U5C output is low or U5D output is low. The outputs of these gates drive the 8-bit digital input and outputs ICs U2 and U3, respectively.

ICs U2 and U3 provide the eight bits of digital input and output, respectively. U2 is a buffer IC with enable input OE*. U3 is an 8-bit register with clock input. It also has a reset input, which I have tied to +5V. The output of U2 is read by the data bus buffer IC U1, and the input to the U3 register IC is provided by the data bus buffer.

An interesting modification to this circuit is to connect the eight digital output pins to the eight digital input pins (i.e., DOUT_0 to DIN_0, DOUT_1 to DIN_1, etc.). With this modification, the digital input lines could be used to read back the data on the digital output lines. Such an output port is called a read-back port. The output ports on the parallel printer port are read-back ports.

Hardware Interrupts

All CPUs have had some sort of interrupt structure built into the CPU chip. The 8088 chip (and later PC chips) have hardware as well as software interrupt features. In all, the 8088 CPU can process 256 interrupt sources. I will consider hardware interrupts in this section and software interrupts in the next section. In the context of programming and using the parallel port, interrupts are very useful because the parallel port has one of the available hardware interrupts, which you can access through the output connector.

Interrupts in a PC are generated either by the external hardware or as the result of a mathematical exception. The hardware interrupts in a PC are handled by the 8259

Programmable Interrupt Controller (PIC) IC by Intel. In a PC, an 8259 provides the capability of eight hardware interrupts to I/O devices. In an AT, an additional 8259 provides a total of 15 interrupts.

Interrupts are generated when an I/O device needs the attention of the CPU. The device signals the PIC, which in turn generates a interrupt to the CPU. The CPU suspends program execution after finishing the current instruction. It then saves the current instruction pointer register onto the stack and proceeds to execute the Interrupt SubRoutine (ISR). After the ISR is completed, the CPU reloads the program counter from the stack (to which it was saved when the interrupt occurred) and restarts the suspended program execution.

On the PC, the 8259-PIC allows eight interrupt channels. Table A.2 shows the hardware interrupt sources and their uses.

BIOS and DOS Interrupts

The software interrupts of the PC are classified as either BIOS or DOS interrupts. BIOS interrupts are universal to all PCs and point to subroutines inside the PC ROM BIOS to provide uniform access to the system hardware for the user programs. These interrupts are built into the PC system.

DOS interrupts, which are also software interrupts, are not built into the PC system; they are provided by the operating system of the PC. These interrupts and corresponding subroutines are loaded into the computer after the operating system has been loaded. Usually, these interrupt routines are used by internal DOS programs.

Table A.2   Hardware interrupts on the PC and PC/AT.

Interrupt Name	Use
NMI	Parity error
IRQ0	System timer
IRQ1	Keyboard
IRQ2	Free on PC, cascaded to IRQ8-15 on AT
IRQ3	COM1
IRQ4	COM2
IRQ5	Hard disk (on XT and AT)
IRQ6	Floppy disk
IRQ7	LPT1
IRQ8 (AT only)	Real-time clock (RTC)
IRQ9 (AT only)	Free
IRQ10 (AT only)	Free
IRQ11 (AT only)	Free
IRQ12 (AT only)	Free
IRQ13 (AT only)	Numeric coprocessor
IRQ14 (AT only)	Hard disk
IRQ15 (AT only)	Free

Appendix B

References

Books

Eggebrecht, Lewis C. Interfacing to the IBM PC. Howard W. Sams and Co.

Green, D. C. Digital Electronic Technology, 2nd Edition. Longman Scientific and Technical.

Horowitz, Paul and Winfield Hill. The Art of Electronics, 2nd Edition. Cambridge University Press.

Messmer, Hans-Peter. The Indispensable PC Hardware Book, 2nd Edition. Addison Wesley.

Osborne, Adam and Gerry Kane. Osborne 16 Bit Microprocessor Handbook. Osborne/McGraw Hill.

Royer, Jeffery P. Handbook of Software and Hardware Interfacing for IBM PCs. Prentice Hall Inc.

Articles

Gadre, Dhananjay V. PC counter uses parallel printer port. *Electronics World + Wireless World* December 1993.

Gadre, Dhananjay V. The parallel printer adapter finds new uses as an instrument interface. *EDN* June 22, 1995.

Gadre, Dhananjay V. A `zero power' ADC for the IBM PC. *Design Ideas. EDN* July 6, 1995.

Gadre, Dhananjay V. Multi channel 12-bit ADC connects to PC. *Design Ideas, EDN* April 25, 1996.

Gadre, Dhananjay V. The parallel adapter as a host interface port. *Dr. Dobb's Journal* April 1996.

Gadre, Dhananjay V. Atmel's AT89C2051 microcontroller. *Dr. Dobb's Journal* July 1997.

Gadre, Dhananjay V. and Larry A. Stein. The enhanced parallel port. *Dr. Dobb's Journal* October 1997.

Gadre, Dhananjay V. and Sunu Engineer. "A data acquisition system for Linux." *Dr. Dobb's Journal*, February 1998.

Gadre, Dhananjay V., P. K. Upadhyay, and V. S. Varma. Catching the right bus, part 2: Using a parallel printer adapter as an inexpensive interface. *Computers in Physics* 8(1), Jan/Feb 1994.

Sunu Engineer and Dhananjay V. Gadre. Data acquisition and distribution in a networked environment. *Embedded Systems Programming* March 1997.

Index

Numerics

A

B

C

D

DAC 79, 80–90
DAC08 85
darlington array driver 46
data acquisition 1–4
data acquisition system 2–3
DATA port 27, 29–31, 32, 33, 36
DATA port address 29
DATA port output details 30
data transfer from a PC to external 55
data transfer overheads 53–57
decoder 23
developing applications using microprocessors 181–183
device driver 257
device driver structure 260–261
digital IC families 15–16
digital signal 8–11, 12–16, 19, 20–22
DIP switch 51

E

ECP 60–61, 74–76
ECR mode for FDC37C665/66 75
ECR mode for PC87332 75
electrical interface for IEEE1284 paral-
  lel ports 59–61, 76–77
emulator 6
EPP 60–76
EPP address read cycle 61, 65
EPP address write cycle 61, 64
EPP BIOS calls 67–68, 69, 71
EPP data read cycle 61, 65
EPP data write cycle 61, 63
EPP port map 69
EPP registers 64–67
EPP signal definitions 62
EPROM 179
EPROM emulator 6, 182–186, 190
EPROM eraser 179
EPROM programmer 6, 223–227

EPROM programmer timing diagram
  225
error signal 26–27
expansion using EPP 163–164
expansion using SPP 158–163
Extended Configuration Register 69

F

FIFO-based waveform generator 255–
  256
finding port addresses 39, 41
flash ADC 91–93
flip-flop 22
fops 258
frequency counter 107–114

H

hardware interfacing 7–9, 13–15
host interface port 206, 212

I

IEEE 1284 standard 60–61
In-Circuit Emulator 181
integrating ADC 94–96
interfacing 7–11
interrupt latency on PCs 34
interrupt subRoutine 8
interrupts 8
IRQ EN signal 31, 33
ISA, PCI, EISA 10

L

latch 14, 22
linking device driver into kernel 261–
  267
Linux 257
LM335 zener 116

M

MAX111 96, 99–100, 102, 103
MAX158 96, 98
MAX158 timing characteristics 127–131
MAX186 96, 98–99
MAX186 ADC signals 139
MAX186 ADC timing diagram 141
MAX186 block diagram 138
MAX186 control byte format 140–142
MAX186 simple interface 142
MAX521 85, 90
MAX521 address byte 146–147
MAX521 block diagram 143
MAX521 command byte 145–147
MAX521 data transfer 144–147
MAX521 signal description 145
microntroller programmer 227–231
Mode 0 operation of 8255-PPIO 166
Mode 1 operation of 8255-PPIO 166
Mode 2 operation of 8255-PPIO 167
MOFSETIRF610 50
MOSFET driver 45, 47–50
MOSFET gate current 48
MOSFET input characteristics 48
MOSFET On voltage 48
MOSFET speed 45, 48
multiplexer 23
multiplying DAC 85

N

Network Printing Alliance 59
nibble mode 60, 61

P

parallel port bit expansion 157–173
parallel printer port anatomy 26–29
PC data area 39–40
PC to printer data transfer 36
period counter 109–114

plug-in interface card disadvantages 1–2
port 7, 10, 11
power switching circuits 45–50
printer cable 31
printer port block diagram 27
programmer 2, 6
PWM DAC 85

R

r-2r ladder DAC 83–84
read-back port 29
reading DIP switches through parallel port 51
reading external data using the STATUS port 55, 57
record temperature 1–4
relay control 45, 46, 47
RS-232 7, 10
RS-232 powered +5V supply 116
Run Length Encoding 75, 76

S

sampling ADC 92–94
scaled-resistance DAC 82
shutter control with MOFSET switch 50
shutter control with MOSFET switch 49
signal convention 14
SmartRAM EPROM emulator 184–185
software driver 7
software interfacing 7–9
speech digitizer 8–11
speed of PC 20
STATUS port 27, 28, 31, 33–35
STATUS port address 29
STATUS port output details 35
strobe signal 26–27, 34
successive approximation principle 97

T

testing hardware on Linux 258–260
time base generator 108–111, 114–115
timing diagram convention 14
transceiver IC 21
TTL and Variants 17
TTL characteristics 17
TTL gates 18, 21
typical signal of sampling ADC 92

U

ULN2003A driver 45–47
unconventional waveform generator 256

W

waveform generator 249–252
weather information on WWW 271–279
wrap-back port 33